The Univ... UNIVERSITY
...d of loa...
AF840544
MAIN LIBRARY WHITEKNIGHTS
90 0428138 6

The normal period of

TELEPEN

90 0428138 6

Techniques of Flavonoid Identification

Biological Techniques Series

J. E. TREHERNE
Department of Zoology
University of Cambridge
England

P. H. RUBERY
Department of Biochemistry
University of Cambridge
England

Ion-sensitive Intracellular Microelectrodes, *R. C. Thomas* 1978
Time-lapse Cinemicroscopy, *P. N. Riddle* 1979
Immunochemical Methods in the Biological Sciences: Enzymes and Proteins, *R. J. Mayer* and *J. H. Waller* 1980
Microclimate Measurement for Ecologists, *D. M. Unwin* 1980
Whole-body Autoradiography, *C. G. Curtis, S. A. M. Cross, R. J. McCulloch* and *G. M. Powell* 1981
Microelectrode Methods for Intracellular Recording and Ionophoresis, *R. D. Purves* 1981
Techniques of Flavonoid Identification. *K. R. Markham* 1982

In preparation
Red Cell Membranes—A Methodological Approach, *J. C. Ellory* and *J. D. Young*

Techniques of Flavonoid Identification

K. R. Markham

Chemistry Division
Department of Scientific and Industrial Research
Petone, New Zealand

547.7
MAR

1982

ACADEMIC PRESS
A Subsidiary of Harcourt Brace Jovanovich, Publishers

London New York
Paris San Diego San Francisco
São Paulo Sydney Tokyo Toronto

391

ACADEMIC PRESS INC. (LONDON) LTD
24–28 Oval Road,
London NW1

U.S. Edition published by
ACADEMIC PRESS INC.
111 Fifth Avenue
New York, New York 10003

Copyright © 1982 by ACADEMIC PRESS INC. (LONDON) LTD

All Rights Reserved
No part of this book may be reproduced in any form by photostat, microfilm, or any other means, without written permission from the publishers

British Library Cataloguing in Publication Data

Markham, K.R.
Techniques of flavonoid identification.—(Biological techniques series)
1. Flavonoids
I. Title
547.7 QP925.F5

ISBN 0-12-472680-1
LCCCN 81-68987

WITHDRAWN
University of
Reading Library

2nd label
21 loans '88

429,138

Typeset by Latimer Trend & Company Ltd, Plymouth
Printed in Great Britain by
St Edmundsbury Press, Bury St Edmunds, Suffolk

To the girls in my life, my wife Pauline
and my daughters, Pamela and Carolyn

Preface

A great deal has already been written on flavonoids, but with few exceptions these are review articles for the specialist which do not discuss practical aspects of techniques in any detail, and so are of limited usefulness to scientists from other disciplines. In the present volume I have attempted to provide the non-specialist with an introduction to, and practical details of, the techniques commonly used to isolate and identify flavonoids from natural sources. However, I am hopeful that much of the information contained herein will also be of value to active researchers in the field, especially if their major discipline is biological rather than chemical. With this in view I have included liberal referencing to alternative techniques and to sources of additional data, which I hope will extend the usefulness of the book as a primary source of information.

The chapters are ordered in a sequence which I feel is best followed when first approaching the problem of flavonoid identification. Thus, isolation and purification of the flavonoid must be achieved first. Information gained from chromatographic behaviour can then be used in conjunction with u.v.-visible absorption data to help identify possible structures. Various forms of hydrolysis may then be appropriate, followed by analysis of the products. The more intractable problems may require additionally the use of more sophisticated techniques such as chemical manipulation, n.m.r. or m.s. Finally, direct comparison with an authentic sample (if possible) is always desirable for confirmation of the proposed structure.

The information presented in this book represents knowledge accumulated over many years of experience in this field, and I am particularly indebted to Professors Tom Mabry (Botany Department, University of Texas) and Jeffrey Harborne (Botany Department, University of Reading) in whose laboratories I gained much of this experience. I am also grateful to my friend and colleague, Dr Lawrence Porter (Chemistry Division, DSIR, New Zealand) whose cooperation and enthusiasm I have enjoyed over the past decade. The support and encouragement I have received from both the administration and staff of the Chemistry Division, DSIR, is also gratefully acknowledged.

The study of flavonoid chemistry and its application to such diverse fields as plant taxonomy and evolution, plant dispersal, plant breeding and fruit and vegetable preservation and processing etc., has proved both challenging and rewarding to me and has added immeasurably to the enjoyment of my scientific career to date. It is my earnest hope that application of the information contained in this book will assist others to gain similarly from this fascinating field of endeavour.

October 1981 K. R. MARKHAM

Contents

1

Introduction to the Flavonoids

It is estimated that about 2% of all carbon photosynthesized by plants (or about 1×10^9 tons per annum) is converted into flavonoids or closely related compounds (Smith, 1972). Most tannins too are flavonoid derived. Flavonoids thus constitute one of the largest groups of naturally occurring phenols. They are virtually ubiquitous in green plants and as such are likely to be encountered in any work involving plant extracts. For this reason it is important that chemists, biochemists, plant physiologists and biologists generally, know how to recognize, isolate and identify these natural products in all their many forms. The following discussion is designed to provide a basic introduction to the flavonoids for newcomers to this field.

1.1 Flavonoid structure variation—general

In plants, flavonoid aglycones (i.e. flavonoids without attached sugars) occur in a variety of structural forms. All contain fifteen carbon atoms in their basic nucleus and these are arranged in a C_6–C_3–C_6 configuration, that is, two aromatic rings linked by a three carbon unit which may or may not form a third ring. For convenience the rings are labelled A, B and C and the individual carbon atoms are referred to by a numbering system which utilizes ordinary numerals for the A- and C-rings and "primed" numerals for the B-ring (see (**1**) but note modified numbering systems used for chalcones, Fig. 1.1).

The flavonoid variants are all related by a common biosynthetic pathway which incorporates precursors from both the "Shikimate" and "Acetate-Malonate" pathways (Hahlbrock and Grisebach, 1975; Wong, 1976), the first flavonoid being produced immediately following confluence of the two pathways (Fig. 1.1). The flavonoid initially formed in the biosynthesis is now

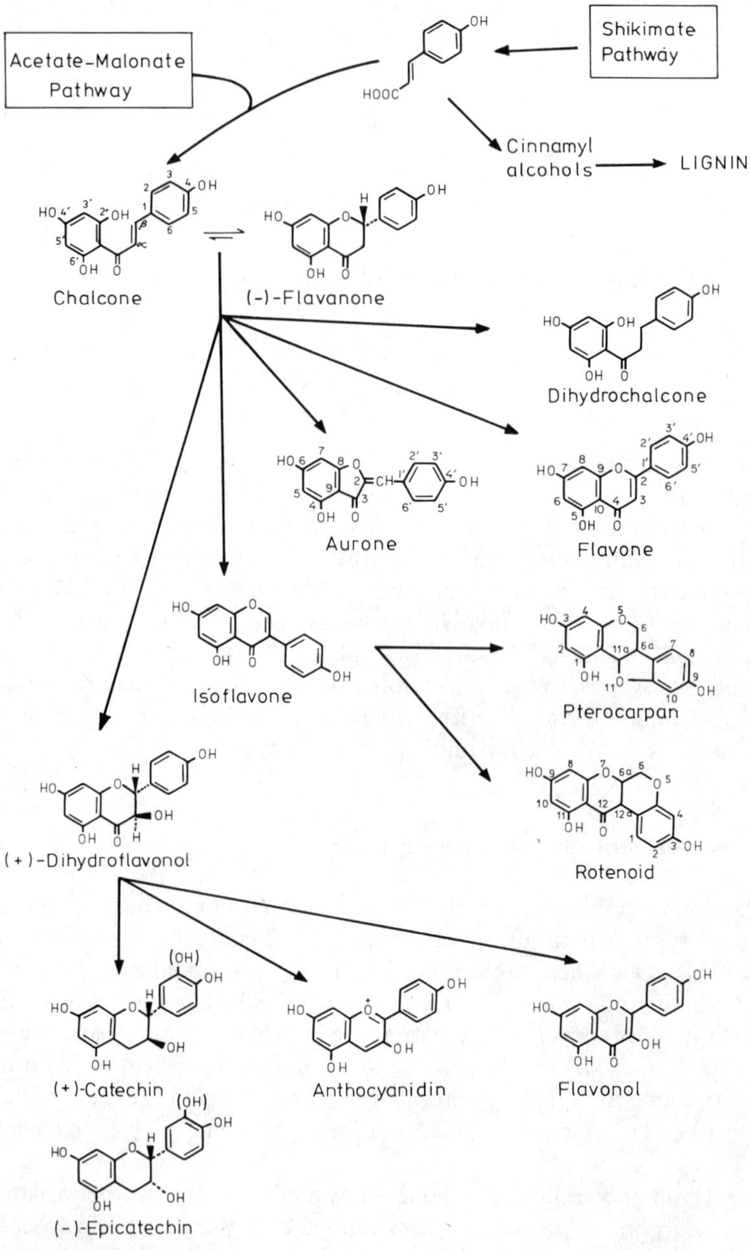

Fig. 1.1. Currently proposed interrelationships between flavonoid monomer types (supported by varying levels of experimental evidence, see Wong, 1976).

thought to be the chalcone (Hahlbrock, 1980) and all other forms are derived from this by a variety of routes (Fig. 1.1). Further modification of the flavonoid may occur at various stages resulting in: additional (or reduced) hydroxylation; methylation of hydroxyl groups or of the flavonoid nucleus;

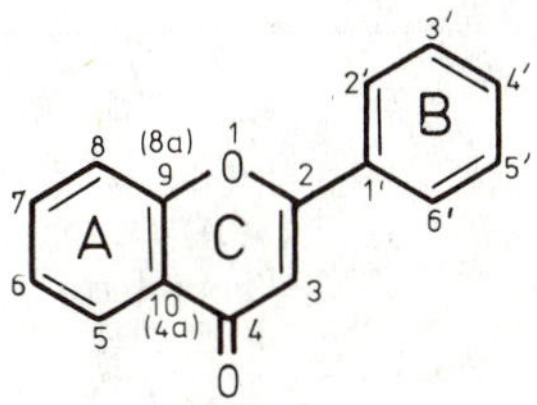

(1)

isoprenylation of hydroxyl groups or of the flavonoid nucleus; methylenation of *ortho*-dihydroxyl groups; dimerization (to produce biflavonoids); bisuphate formation; and most importantly, glycosylation of hydroxyl groups (to produce *flavonoid O-glycosides*) or of the flavonoid nucleus (to produce *flavonoid C-glycosides*). The range of known flavonoids is thus vast and lists of known variants have been published (Harborne *et al.*, 1975) and recently updated (Harborne and Mabry, 1982). For the reader unfamiliar with the myriad of "trivial" names given to many flavonoids, an excellent summary has been compiled by Swain (1976) which relates name, structure and primary source and another by Wollenweber and Dietz (1981). A selection of these from Swain (1976) and Geissman (1962) is included here in Table 1.1 as an aid to understanding later chapters in this book in which trivial names are used for both convenience and conciseness.

Table 1.1
A selection of frequently encountered flavonoid aglycones, their trivial names, structures and primary sources

Flavonoid aglycones	Structure	Source
Flavones		
Chrysin	5,7-OH	*Populus*
Baicalein	5,6,7-OH	*Scutellaria*
Apigenin	5,7,4′-OH	*Petroselinum*
Acacetin	4′-Me apigenin	*Robinia*
Scutellarein	5,6,7,4′-OH	*Scutellaria*
Hispidulin	6-Me scutellarein	*Ambrosia*
Luteolin	5,7,3′,4′-OH	*Reseda*
Chrysoeriol	3′-Me luteolin	*Eriodictyon*
Diosmetin	4′-Me luteolin	*Diosma*
Tricetin	5,7,3′,4′,5′-OH	*Lathyrus*
Tricin	3′,5′,-Me tricetin	*Triticum*

Table 1.1 (contd)

Flavonoid aglycones	Structure	Source
Flavonols		
Galangin	3,5,7-OH	*Alpinia*
Fisetin	3,7,3′,4′-OH	*Rhus*
Kaempferol	3,5,7,4′-OH	*Delphinium*
Kaempferide	4′-Me kaempferol	*Alpina*
Robinetin	3,7,3′, 4′, 5′-OH	*Robina*
Herbacetin	3,5,7,8,4′-OH	*Gossypium*
Quercetin	3,5,7,3′,4′-OH	*Quercus*
Rhamnetin	7-Me quercetin	*Rhamnus*
Isorhamnetin	3′-Me quercetin	*Cheiranthus*
Myricetin	3,5,7,3′,4′,5′-OH	*Myrica*
Quercetagetin	3,5,6,7,3′,4′-OH	*Tagetes*
Gossypetin	3,5,7,8,3′,4′-OH	*Gossypium*
Anthocyanidins		
Apigenidin	5,7,4′-OH	*Rechsteineria*
Luteolinidin	5,7,3′,4′-OH	*Rechsteineria*
Pelargonidin	3,5,7,4′-OH	*Pelargonium*
Cyanidin	3,5,7,3′,4′-OH	*Centaurea*
Peonidin	3′-Me cyanidin	*Paeonia*
Delphinidin	3,5,7,3′,4′,5′-OH	*Delphinium*
Petunidin	3′-Me delphinidin	*Petunia*
Malvidin	3′,5′-Me delphinidin	*Malva*
Isoflavones		
Daidzein	7,4′-OH	*Pueraria*
Formononetin	4′-Me daidzein	*Ononis*
Genistein	5,7,4′-OH	*Genista*
Biochanin-A	4′-Me genistein	*Cicer*
Orobol	5,7,3′,4′-OH	*Orobus*
Tectorigenin	5,7,4′-OH 6-OMe	*Iris*
Baptigenin	5,7,3′,4′,5′-OH	*Baptisia*
Flavanones		
Pinocembrin	5,7-OH	*Pinus*
Liquiritigenin	7,4′-OH	*Glycyrrhiza*
Naringenin	5,7,4′-OH	*Prunus*
Sakuranetin	7-Me naringenin	*Prunus*
Eriodictyol	5,7,3′,4′,-OH	*Eriodictyon*
Hesperetin	4′-Me eriodictyol	*Prunus*

Flavonoid aglycones	Structure	Source
Dihydroflavonols		
Pinobanksin	3,5,7-OH	*Pinus*
Aromadendron	3,5,7,4′-OH	*Eucalyptus*
Fustin	3,7,3′,4′-OH	*Rhus*
Taxifolin	3,5,7,3′,4′-OH	*Pseudotsuga*
Biflavonoids		
Agathisflavone	6,8″-biapigenin	*Agathis*
Cupressuflavone	8,8″-biapigenin	*Cupressus*
Amentoflavone	3′,8″-biapigenin	*Cupressus*
Ginkgetin	amentoflavone 7,4′-dimethyl ether	*Ginkgo*
Sciadopitysin	amentoflavone 7,4′,4‴ trimethyl ether	*Ginkgo*
Robustaflavone	6,3‴-biapigenin	*Agathis*
Hinokiflavone	6,4‴-bi-O-apigenin	*Cupressus*
Ochnaflavone	3′,4‴-bi-O-apigenin	*Ochna*
Chalcones[a]		
Isoliquiritigenin	2′,4′,4-OH	*Acacia*
Chalconaringenin	2′,4′,6′,4-OH	*Salix* (as 2′-*O*-glucoside)
Butein	2′,4′,3,4-OH	*Acacia*
Okanin	2′,3′,4′,3,4,-OH	*Acacia*
Aurones[a]		
Sulphuretin	6,3′,4′-OH	*Bidens*
Aureusidin	4,6,3′,4′-OH	*Antirrhinum*
Maritimetin	6,7,3′,4′,-OH	*Bidens*
Leptosidin	6,3′,4′,-OH,7-OMe	*Coreopsis*

[a] Note different numbering systems used (Fig. 1.1).

1.2 Flavonoid *O*-glycosides

Flavonoids commonly occur as flavonoid *O*-glycosides in which one or more of the flavonoid hydroxyl groups is bound to a sugar or sugars by an acid-labile hemiacetal bond (e.g. (**2**)). The effect of glycosylation is to render the flavonoid less reactive and more water (sap) soluble, the latter property permitting storage of the flavonoids in the cell vacuole (where they are commonly found). Although hydroxyl groups in any position on the flavonoid nucleus may be glycosylated, in fact hydroxyls in certain sites have a much higher probability of being so than others, e.g. the 7-hydroxyl in flavones, isoflavones and dihydroflavones, the 3- (and 7-) hydroxyl in flavonols and dihydroflavonols, and the 3- (and 5-) hydroxyl in anthocyanidins. Glucose is the sugar most commonly involved, although galactose, rhamnose, xylose and arabinose are not uncommon. Other sugars occasionally encountered include allose, mannose, fructose, apiose and glucuronic and galacturonic acids. Disaccharides are also often found in association with flavonoids, e.g. sophorose (2-*O*-β-D-glucosyl-D-glucose), gentiobiose (6-*O*-β-D-glucosyl-D-glucose), rutinose (6-*O*-α-L-rhamnosyl-D-glucose) and neohesperidose (2-*O*-α-L-rhamnosyl-D-glucose), and occasionally tri- and even tetra-saccharides. It is accepted that in plants, *O*-glycosylation (and methylation) occurs as one of the last stages in the biosynthesis and is catalysed by enzymes of high specificity. Glycosides occasionally exhibit one further modification, that of acylation. Acylated glycosides have one (or more) of their sugar hydroxyls derivatized with an acid such as acetic or ferulic. The bond in this case is an ester bond, the acid effectively being esterified by the sugar, as for example in (**3**). The range of flavonoid *O*-glycosides found in nature has been summarized (Harborne *et al*., 1975) and recently updated (Harborne and Mabry, 1982).

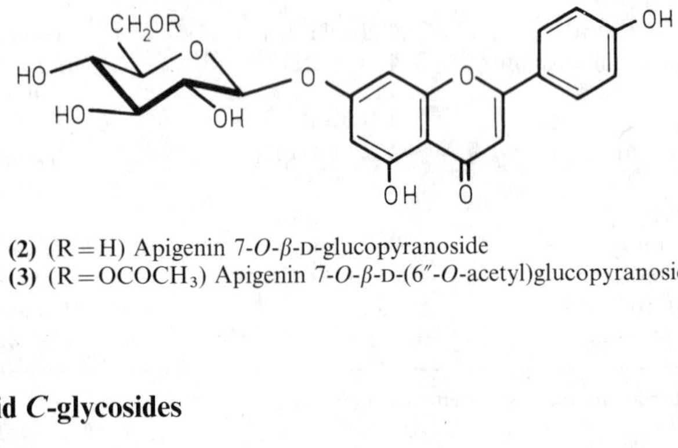

(2) (R=H) Apigenin 7-*O*-β-D-glucopyranoside
(3) ($R=OCOCH_3$) Apigenin 7-*O*-β-D-(6″-*O*-acetyl)glucopyranoside

1.3 Flavonoid *C*-glycosides

Sugars may also be *C*-linked to the flavonoid and in this case they are attached directly to the benzene nucleus by a carbon-carbon bond (e.g. (**4**)) which is acid resistant (c.f. *O*-glycosides). Such glycosides are referred to as flavonoid *C*-glycosides. To date *C*-linked sugars have been found only at the 6- and 8-positions on flavonoid nuclei. The range of sugars involved is apparently very much smaller than in *O*-glycosides and includes glucose most commonly (e.g. vitexin, orientin), and also galactose (e.g. apigenin 8-*C*-galactoside), rhamnose (e.g. violanthin), xylose (e.g. vicenin-1) and arabinose (e.g. schaftoside). The range of flavonoid aglycone types involved is also very restricted. Thus, although isoflavones, flavanones and flavonols occur occasionally in *C*-glycosylated form, flavone *C*-glycosides are by far the most prevalent. As with *O*-glycosides, *C*-glycosides are often found

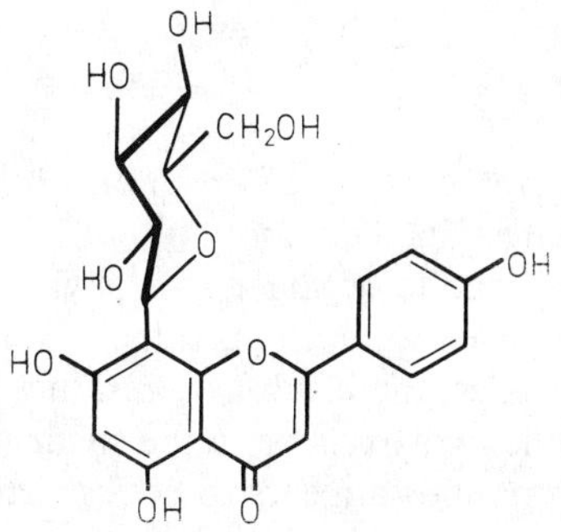

(4) Apigenin 8-*C*-β-D-glucopyranoside (vitexin)

further derivatized by *O*-glycosylation (at the sugar or phenolic hydroxyls) or by acylation (normally of a sugar hydroxyl). Comprehensive lists of known flavonoid *C*-glycosides have recently been published (Chopin and Bouillant, 1975; Chopin *et al.*, 1982), and a selection is presented here in Table 1.2.

Table 1.2
The trivial names and structures of selected naturally occurring flavonoid *C*-glycosides

Trivial name	Aglycone	*C*-linked sugars C-6	*C*-linked sugars C-8
Vitexin	Apigenin	—	Glucose
Isovitexin	Apigenin	Glucose	—
Vicenin-1	Apigenin	Xylose	Glucose
Vicenin-2	Apigenin	Glucose	Glucose
Vicenin-3	Apigenin	Glucose	Xylose
Violanthin[a]	Apigenin	Glucose	Rhamnose
Schaftoside[a]	Apigenin	Glucose	Arabinose
Orientin	Luteolin	—	Glucose
Isoorientin	Luteolin	Glucose	—
Scoparin[a]	Chrysoeriol	—	Glucose
Keyakinin	7-Methyl kaempferol	Glucose	—
Hemiphloin[a]	Naringenin	Glucose	—
Puerarin	Daidzein	Glucose	—
Paniculatin	Genistein	Glucose	Glucose

[a] "Iso" forms in which the sugar position (s) is (are) reversed are also naturally occurring.

1.4 Flavonoid sulphates

The only other water-soluble group of flavonoids likely to be encountered are the flavonoid sulphates. Such compounds contain one or more sulphate residues attached to a phenolic or sugar hydroxyl. Technically these compounds are bisulphates since they occur as salts, i.e. flavone-*O*-SO_3K. Many are bisulphates of glycosides, the bisulphate residues being attached to any of the remaining phenolic hydroxyls or to the sugar (Harborne, 1977). The occurrence of these compounds appears to be restricted to the angiosperms and especially to those angiosperms which have an ecological association with aquatic habitats (Harborne, 1975).

1.5 Biflavonoids

As the name implies, biflavonoids are flavonoid dimers, although the procyanidin dimers (catechin based) are normally excluded from this group. The flavonoids commonly involved are flavones and flavanones with the biosynthetically simple 5,7,4′ (or occasionally, 5,7,3′,4′) oxygenation pattern, the interflavonoid linkage being either a carbon–carbon bond or (occasionally) an ether bond (Geiger and Quinn, 1975). Flavonoid monomers incorporated in each biflavonoid may be of the same or of different type and linkage positions vary widely. The more common carbon–carbon linked types include: 6,8″-linked (agathisflavone group), 8,8″-linked (cupressuflavone group), 3′,8″-linked (amentoflavone group, e.g. (**5**)), 6,3‴-linked (robustaflavone group), and 3,8″-linked. The ether linked types include: 6,4‴-linked (hinokiflavone group, e.g. (**6**)) and 3′, 4‴-linked (ochnaflavone group), see Table 1.1.

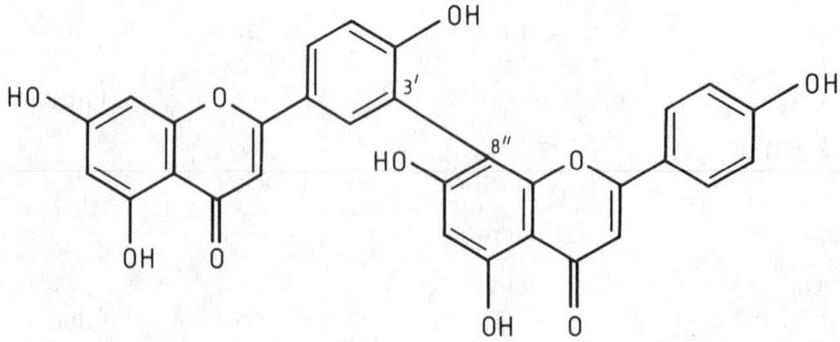

(**5**) Amentoflavone

Many physical and chemical properties of biflavonoids resemble those of their component monoflavonoids (e.g. u.v.-visible spectra, colour tests, etc.) and as a consequence, biflavonoids can sometimes be difficult to recognize. However, chromatography on silica distinguishes monomers and dimers clearly (Chexal *et al.*, 1970) and confirmation may be achieved by the use of alkaline fusion or by mass spectroscopy.

Biflavonoids have rarely been found as glycosides and have a restricted distribution (see Table 1.3), occurring predominantly in the Gymnosperms. A comprehensive review article detailing their natural occurrence and structural variation has appeared (Geiger and Quinn, 1975) and has recently been updated (Geiger, 1982).

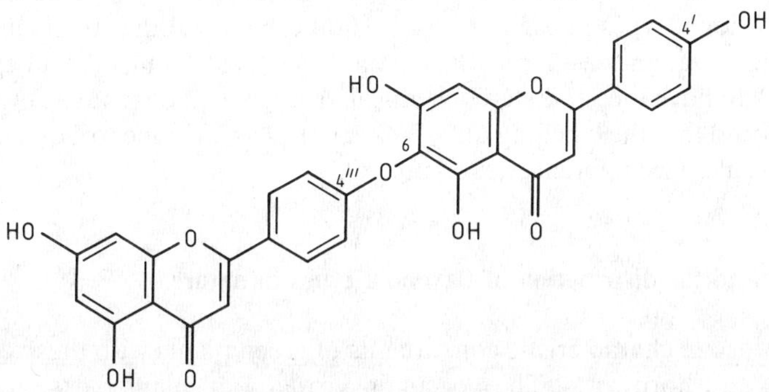

(6) Hinokiflavone

1.6 Optically active flavonoid aglycones

As is evident from the structures in Fig. 1.1, a number of flavonoid aglycone types possess asymmetric carbon atoms and thus exhibit optical activity (i.e. they rotate plane-polarized light). Included in this class of flavonoids are flavanones, dihydroflavonols, catechins, pterocarpans, rotenoids and some biflavonoids. The rotation (by convention at 589·3 nm, the sodium D-line) of a naturally occurring flavonoid aglycone is related to the absolute stereochemistry of the flavonoid. Thus, the normal, laevorotatory (−)-flavanones have the "S" configuration at C-2, i.e. possess 2S configuration, whereas the (+)-flavanones possess 2R configuration, in which the proton and B-ring are interchanged (see Cahn and Ingold, 1951; Cahn *et al.*, 1956, for details of nomenclature and chirality rules).

A list of useful rotation-structure correlations is presented below (*cis*- and *trans*- nomenclature refers to the relationship between the protons on adjacent asymmetric carbon atoms):

(−) Flavanones	2S	(Bohm, 1975)
(+) Dihydroflavonols (*trans*)	2R:3R	(Bohm, 1975)
(−) Epicatechin (*cis*)	2R:3R	(Haslam, 1975)
(+) Epicatechin (*cis*)	2S:3S	(Haslam, 1975)
(+) Catechin (*trans*)	2R:3S	(Haslam, 1975)
(−) Pterocarpans (*cis*)	6aR:11aR	(Ingham and Markham, 1980)
(−) 6a-Hydroxypterocarpans (*cis*)	6aS:11aS	(Ingham and Markham, 1980)
(+) 6a-Hydroxypterocarpans (*cis*)	6aR:11aR	(Ingham and Markham, 1980)

Apart from (+)-epicatechin, the above are the commonly encountered configurational types in each group. It should be noted that the commonly occurring flavanones (−), dihydroflavonols (+), catechin (+) and epicatechin (−) all have the *same* stereochemical arrangement of groups at C-2 (see Fig. 1.1) and that the labelling of (−)-flavanones as 2S is due to a quirk of the chirality rules (see references cited above).

1.7 Guide to the distribution of flavonoid types in nature

Flavonoids are characteristic constituents of green plants with the exception of algae and hornworts. They occur in virtually all plant parts including leaves, roots, wood, bark, pollen, nectar, flowers, berries and seeds. In the few recorded cases of flavonoids being found in animals, for example in the beaver scent gland, propolis (bee secretion) and in butterfly wings, it is considered that the flavonoids originate from the plants upon which the animals feed rather than being biosynthesized *in situ* (Harborne, 1967). The distribution of flavonoids in the plant kingdom (as revealed to date) is presented in Table 1.3 as a guide to the likely flavonoid types to be encountered in investigations of particular plant groups. The distribution of flavonoid types within the largest group, the angiosperms, is further elaborated in a recent review by Gornall *et al.* (1979). It is evident from the data in this Table that apart from the single example of flavonoids being found in the Chlorophyta (Markham and Porter, 1969), and the confirmed synthesis of a halogenated flavone by a fungus (Marchelli and Vining, 1973), flavonoids are restricted to plant groups with a degree of complexity of the Bryophyta or higher. Perhaps significantly, these same groups are all of more recent origin (up to 460 million years) than are the groups lacking flavonoids (500–3000 million years). A general trend evident from this Table is the increased range of flavonoids present in the more highly evolved plant groups.

Another important feature of the distribution of flavonoids in plants is the

Table 1.3
Distribution of flavonoids in the plant kingdom

Divisions	Common name[a]	Flavonoids found[b]
Procaryote		
Schizophyta	Bacteria (3200?)	None
Cyanophyta	Blue-green algae (2800)	None
Eucaryote		
Chlorophyta	Green algae (1000)	None
	Stoneworts (450) (Charophyceae)	Flavone *C*-glycosides (rare. *Nitella*)
Phaeophyta	Brown algae (400–600)	None
Rhodophyta	Red algae (600–900)	None
Fungi	Fungi (460)	Dihydrochalcone (rare. *Phallus*); unusual flavone (rare. *Aspergillus*)
	(Lichens)	None
Bryophyta	Liverworts (380)	[c]Flavone *C*- and *O*-glycosides
		Flavonol *O*-glycosides (rare)
		Flavanone *O*-glycoside (rare. *Riccia*)
		Aurone *O*-glycoside (rare)
		Dihydrochalcone *C*-glycoside (rare? *Hymenophyton*)
	Mosses (280)	(Flavonol *O*-glycosides?)
		Biflavones (rare. *Dicranum*)
		Aurone (rare. *Funaria*)
		[c]Flavone *C*- and *O*-glycosides
		3-Deoxyanthocyanins
	Hornworts (?) (Anthocerotae)	None
Tracheophyta	Psilophytes (400)	[c]Biflavones, *O*-glycosides
		Flavones, *O*-glycosides (trace)
		Flavone *C*-glycosides (trace)
	Lycopods (380)	[c]Flavones, *O*-glycosides
		Flavones, *C*-glycosides (rare. *L. cernuum*)
		Biflavones (*Selaginella* only)
	Horsetails (380)	[c]Flavonol *O*-glycosides
		Flavone *O*-glycosides
		Proanthocyanidins
		Flavone *C*-glycosides (trace?)
		Flavanones, dihydroflavonols
	Ferns (375)	[c]Flavonol-*O*-glycosides
		Flavones, *C*- and *O*-glycosides (rare)
		3-Deoxyanthocyanins
		Flavanones
		Chalcones, dihydrochalcones
		Biflavones (rare. *Osmunda* only)
		Proanthocyanidins, anthocyanins

Table 1.3 (contd)

Divisions	Common name[a]	Flavonoids found[b]
	Gymnosperms (350)	Flavones Biflavonoids Flavanones Flavone *C*-glycosides Isoflavones (rare, e.g. *Juniperus*) Proanthocyanidins, anthocyanins Flavonols, dihydroflavonols
Angiosperms	Angiosperms (135)	[c]Flavones and flavonols, *C*- and *O*-glycosides and bisulphates Isoflavones, *C*- and *O*-glycosides (Leguminosae) Flavanones, *C*- and *O*-glycosides Chalcones and dihydrochalcones *C*- and *O*-glycosides Proanthocyanidins and anthocyanins Aurones, *O*-glycosides Biflavones (rare) Dihydroflavonols, *O*-glycosides

[a] Figures in parentheses following the common name refer to the age in millions of years of the oldest known fossil (Swain, 1975).
[b] Where *O*-glycosides are mentioned, aglycones are generally also found, Generic names are listed where only one genus appears to be involved.
[c] Commonly encountered.

strong tendency for taxonomically related plants to produce similar types of flavonoids. Thus, useful information about the flavonoid types likely to be encountered in a plant under investigation may often be gained by reference to the literature pertaining to previous flavonoid studies of related plants, e.g. from the same genus or family.

1.8 Suggested stepwise procedure for flavonoid structure determination

The following "key" is presented as a guide for non-specialists in the field of flavonoid identification. Although not exhaustive in its treatment, it does indicate a sensible sequence of procedures which will guide the researcher through to possible structures for most flavonoids, using the techniques described in this book.

1. Be cognizant of flavonoid types normally found in related plants.

2. Purify flavonoid to be studied (Chapter 2).
3. From the mobility of the flavonoid on a paper chromatogram (p.c.) determine whether it is likely to be a glycoside or an aglycone (Chapter 2). If an aglycone, omit steps 4–7.
4. If a glycoside, measure u.v.-visible absorption spectra (Chapter 3) and hydrolyse with acid (Chapter 4).
5. Analyse products by p.c. to determine whether any aglycone is produced and so distinguish *O*-glycosides, *O*-glycuronides and *C*-glycosides, and also acylated and sulphated flavonoids (Chapter 4). If a *C*-glycoside, omit steps 6–17. If acylation suspected, refer to step 16.
6. Separate sugar(s) and aglycone (Chapter 4).
7. Analyse sugar(s) by p.c. or g.l.c. (Chapter 4).
8. Measure u.v.-visible spectra for purified aglycone (Chapter 3) and attempt to interpret, and then to match spectra with those of known flavonoids (Chapter 3).
9. If a possible match is made, compare aglycone directly with an authentic sample (Chapter 7), or confirm structure using mass spectrometry (m.s.), nuclear magnetic resonance (n.m.r.) (Chapter 6), or elemental analysis.
10. If a match is not made use m.s. or n.m.r. techniques to further define the structure (Chapter 6).
11. If the aglycone is new or unobtainable, attempt inter-conversion to a known aglycone, or vice versa (Chapter 5).
12. Once sugar(s) and aglycone are identified, attempt to define site of *O*-glycosylation by comparing the u.v.-visible spectra of aglycone and glycoside (Chapter 3). Alternatively, compare ^{13}C-n.m.r. spectra (Chapter 6).
13. If more than one sugar is indicated by the sugar analysis or by the p.c. mobility of the glycoside (Chapter 2), there may be more than one site of *O*-glycosylation. Therefore, carry out partial and/or enzymic hydrolyses (Chapter 4) and analyse products as outlined above.
14. If two or more sugars are attached to one site, the interglycosidic linkage is best identified by ^{13}C-n.m.r. (or occasionally by ^{1}H-n.m.r. or m.s.) (Chapter 6.)
15. Compare flavonoid glycoside (if previously reported) directly with an authentic sample (Chapter 7), or alternatively compare data with published data. Remember, sulphate and acyl groups may have been missed at step 5! Their presence will affect physical properties.
16. For acyl group identification and positioning, use alkaline hydrolysis (Chapter 4) for cleavage and t.l.c. (Chapter 2), ^{13}C-n.m.r. or m.s. (Chapter 6) for identification and ^{13}C-n.m.r. for positioning.
17. If a C-linked glycoside is indicated at step 5 and no change in mobility was evident on acid treatment, attempt Wessely–Moser rearrangement

(Chapter 4). Mono- and di-*C*-glycosides are distinguished by p.c. mobility. For sugar identification, refer to Chapter 4.

18. If a *C*-linked glycoside is indicated at step 5, and a change in p.c. mobility was evident on acid treatment, the *C*-glycoside is *O*-glycosylated. Determine site of glycosylation as in step 12.

19. If step 18 indicates *O*-glycosylation of the *C*-linked sugar, use m.s. or ^{13}C-n.m.r. data (Chapter 6) to define site. Then refer to step 15.

2

Isolation and Analytical Techniques

2.1 Solubility characteristics of flavonoids

Flavonoid aglycones are polyphenols and as such possess the chemical properties of phenolics. That is, they are slightly acidic and will thus dissolve in alkali. Note however that if left in alkali, in the presence of oxygen, many will degrade (see chapter on u.v.-visible absorption spectra). Flavonoids possessing a number of unsubstituted hydroxyl groups, or a sugar, are polar compounds and as the old adage "like dissolves like" suggests, are generally moderately soluble in polar solvents such as ethanol (EtOH), methanol (MeOH), butanol (BuOH), acetone, dimethyl sulphoxide (DMSO), dimethyl formamide (DMF), water etc. The presence of an attached sugar (commonly encountered) tends to render the flavonoid more water soluble and combinations of the above solvents with water are thus better solvents for glycosides. In contrast, less polar aglycones such as isoflavones, flavanones and highly methoxylated flavones and flavonols tend to be more soluble in solvents such as ether and chloroform.

2.2 Selection, preparation and extraction of plant material

Fresh vegetation provides the ideal starting material for a flavonoid analysis, although dried samples which have been carefully stored for many years may still be satisfactory. Herbarium specimens of over 100 years old have proved suitable for flavonoid analysis (Harborne, 1967b) and flavonoids have even been isolated from fossils 25 million years old (Niklas and Gianassi, 1978). However, there is a tendency in old plant material for glycosides to be converted to aglycones by fungal action and for sensitive aglycones to be oxidized.

With fresh plant material it is recommended that, after a selected sample has been set aside as a voucher specimen, the remainder be dried rapidly (to prevent enzyme action) in an oven at about 100°C. The dried plant material can then either be stored in a sealed plastic bag for future use, or ground to a fine powder for solvent extraction. If the dried plant material does not grind up easily with a mortar and pestle, it may be macerated in a Waringblender with or without the solvent to be used for extraction. It is often a wise precaution to extract a sample of undried plant material for a later chromatographic check that the drying process has not altered the flavonoid composition.

After weighing out a portion of the ground plant material, extraction is best carried out in two steps, firstly with MeOH:H_2O (9:1) and secondly with MeOH:H_2O (1:1). At each step, sufficient solvent is added to make a liquid slurry and the mixture is left for 6–12 hours. Filtration to separate extract from plant material can be carried out rapidly by using a glass wool or cotton wool plug in the neck of a filter funnel, or better, by suction using a Buchner funnel (recommended paper: Whatman No 54 or 541 or equivalent). The two extracts are then combined and evaporated to about $\frac{1}{3}$ the original volume, or until most of the MeOH has been removed. The resultant aqueous extract can then be cleared of low polarity contaminants such as fats, terpenes, chlorophylls, xanthophylls etc. by extraction (in a separating funnel) with hexane or chloroform. This should be done several times and the extracts combined. Although unlikely to contain flavonoids, the combined extract should not be discarded until checked by chromatography. The solvent-extracted aqueous layer, containing the bulk of the flavonoids, is then evaporated to dryness under vacuum in a rotary evaporator.

The above procedure is suitable for the extraction of most flavonoids but not for anthocyanins or for low polarity flavonoids (which sometimes occur on the exterior of the plant). For anthocyanins, fresh leaves or petals should not be dried but should be crushed with MeOH containing 1% concentrated HCl. Extraction is almost immediate as evidenced by the colour in solution, and chromatographic or spectroscopic analyses of the extract should be carried out as soon after extraction as possible in order to minimize the risk of glycoside hydrolysis occurring. The low polarity flavonoids occasionally found on the exterior of desert plants, ferns etc. are best isolated by simply soaking the fresh plant material in hexane or ether for a few minutes. The resultant extract will also contain waxes and fats which can be separated later by chromatography.

2.3 Paper chromatography and the recognition of flavonoids

Paper chromatography (p.c.) is probably the most generally useful chromatographic technique available to flavonoid chemists today. For this reason, the technique will be dealt with in some detail here. Preliminary analysis of a plant extract for the presence of flavonoids is conveniently accomplished by this method, the separation normally being carried out by two-dimensional p.c. (2D p.c.), in the descending mode.

2.3.1 *General technique*

The paper recommended for this purpose is Whatmans 3MM (46 × 57 cm) or equivalent. It should be folded as illustrated in Fig. 2.1(a) to permit the securing of the paper in a trough for descending chromatography. A solution of the plant extract is then applied ("spotted") to the paper at a point about 8 cm in from the side edge and 3 cm in from the last fold (Fig. 2.1(a)). The extract should be applied evenly on to a 3-cm diameter circle centred at this point, drying of the spot being aided by the use of a hairdryer. The amount of extract applied can be a critical factor in determining the quality of the

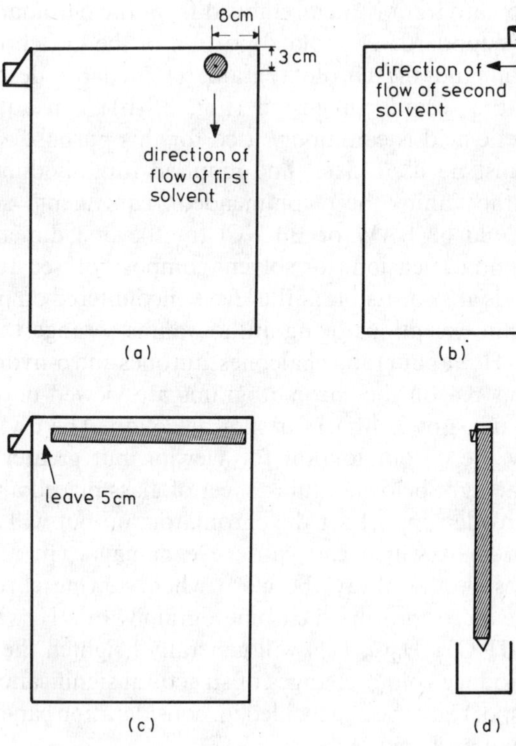

Fig. 2.1. Paper chromatographic methodology.

flavonoid pattern ultimately obtained. If too little extract is applied, the resultant flavonoid spots may be hard to detect, whereas if too much is applied, smearing and poor resolution will result. If in doubt, a good general guide is to apply the amount of extract obtained from 50–100 mg of dry plant material.

Chromatography tanks of a size suitable for these large sheets are commercially available, e.g. Shandon Panglass chromatatanks (all glass, and recommended) and Research Specialties Company (Berkley) chromatocab, or alternatively can be constructed (Mabry *et al.*, 1970). Smaller all-glass tanks may be used if the paper sheets used are halved or quartered before spotting the extract. If this is done, however, the spot size of 3 cm diameter and the level of loading must be reduced accordingly.

For preliminary assessment of the flavonoid content of an extract, it is general practice to run the spotted paper first in an alcoholic solvent such as BAW or TBA (see later Table 2.2) in the longer dimension. Depending upon the room temperature, it will take the solvent front 12–30 hours to approach the end of the sheet (about 50–60 ml of solvent is required per sheet). The chromatogram is then removed from the chromatocab and air-dried in a fume cupboard. The folded portion of the dry chromatogram is then trimmed off and the chromatogram refolded for descending chromatography in the second dimension (Fig. 2.1(b)). An aqueous solvent such as 15% acetic acid is commonly used for this purpose and a separate chromatocab must be used. The time required for this dimension is 4–6 hours. With anthocyanins the recommended "equivalent" solvents (Harborne, 1973) would be BAW or Bu/HCl for the first dimension and 1% HCl for the second dimension (for solvent composition see Table 2.2).

Most flavonoids are not visible at the levels encountered on paper chromatograms, the main exceptions being anthocyanins (orange to purple spots turning blue in NH_3 vapour) and chalcones, aurones and 6-hydroxy flavonols (yellow). For this reason the chromatograms are viewed under ultraviolet (u.v.) light (366 nm, not 254 nm) for spot detection. The u.v. light may be above or below the chromatogram for viewing but greater sensitivity is achieved with the light below. A thick sheet of glass placed above the lamp will not only provide support for the chromatogram but will also filter out high-energy u.v. light which can cause eye damage. (In any case, u.v.-absorbing glasses should always be worn when viewing chromatograms.) Treatment of the *thoroughly* dried chromatogram with NH_3 vapour (from a bottle of 0·88 $NH_4OH:H_2O$, 1:1) will generally heighten the sensitivity of detection and produce colour changes of structural significance (see below). Once all visible spots have been pencilled in, consideration can be given to the chemical nature of each spot.

2.3.2 *Flavonoid structure information from p.c. data*

The fact that a spot is visible at all under these conditions is a likely indication that it represents a phenolic. Flavonoids often account for the majority of visible (in u.v.) spots although blue fluorescent, pink, whitish, orange and brownish spots must be considered as unlikely to represent flavonoids until further investigated (by u.v.-visible spectroscopy). The "typical" flavone and flavonol glycoside spot will appear dark purple in the u.v. and will turn yellow or yellow-green in NH_3 but a number of other colour combinations are also encountered. The range of spot colours which may be associated with flavonoids is detailed in Table 2.1 together with possible flavonoid structure relationships.

Table 2.1
Interpretation of spot colour in terms of flavonoid structure

Spot colour in ultraviolet light: u.v. light without NH_3	Spot colour in ultraviolet light: u.v. light with NH_3	Possible flavonoid type
Dark purple	Yellow, yellow-green or green	a. Commonly 5-OH flavones or flavonols (3-*O*-substituted with 4′-OH b. Occasionally 5-OH flavanones and 4′-OH chalcones with no B-ring -OHs
	Little or no colour change	a. Commonly flavones or 3-*O*-substituted flavonols with 5-OH but lacking a *free* 4′-OH b. Some 6- or 8-OH flavones and 3-*O*-substituted flavonols with 5-OH c. Isoflavones, dihydroflavonols, biflavonyls and some flavanones with 5-OH d. Chalcones with 2′- or 6′-OH but without a free 2- or 4-OH
	Light blue	Some 5-OH flavanones
	Red or orange	Chalcones with a free 2- and/or 4-OH
Fluorescent light blue	Fluorescent yellow-green or fluorescent blue-green	a. Flavones and flavanones lacking a free 5-OH, e.g. 5-*O*-glycosides b. Flavonols lacking a free 5-OH but with the 3-OH substituted
	Little or no colour change	Isoflavones lacking a free 5-OH
	Bright fluorescent light blue	Isoflavones lacking a free 5-OH

Table 2.1 (contd)

Spot colour in ultraviolet light: u.v. light without NH_3	Spot colour in ultraviolet light: u.v. light with NH_3	Possible flavonoid type
Invisible	Fluorescent light-blue	Isoflavones lacking a free 5-OH
Dull yellow and yellow, or orange fluorescence	Little or no colour change	Flavonols with a free 3-OH and with or without a free 5-OH (sometimes originating from the dihydroflavonol)
Fluorescent yellow	Orange or red	Aurones with a free 4′-OH and some 2- or 4-OH chalcones
Yellow-green blue-green or green	Little or no colour change	a. Aurones lacking a free 4′-OH and flavanones lacking a free 5-OH b. Flavonols with a free 3-OH and with or without a free 5-OH
Dull orange red or mauve	Blue	Anthocyanidin 3-glycosides
Cerise pink or fluorescent yellow	Blue	Most Anthocyanidin 3,5-diglycosides

Another valuable source of structure information is the position of the spot on the 2D-p.c. Thus, viewed with the origin spot to the lower right, the chromatogram will display all flavonoid aglycones along the base and left-hand side, and glycosides over the rest of the area. A guide as to where particular flavonoid types might be expected to appear, using the "standard" chromatographic conditions described above, is given in Fig. 2.2 (adapted from Mabry *et al.*, 1970). For equivalent data on anthocyanins see Harborne (1967a).

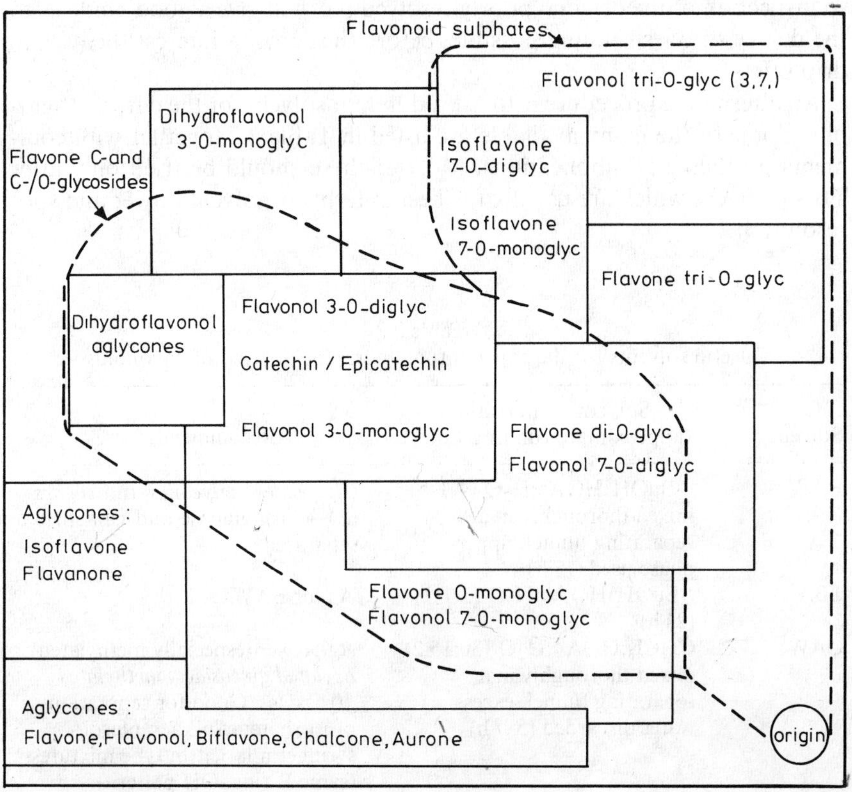

Fig. 2.2. A guide to the distribution of flavonoid types on a chromatogram developed in TBA/15% HOAc.

2.3.3 *Solvents and the improvement of spot resolution*

Occasionally groups of two or more spots on a p.c. will be poorly resolved. If these spots are of low or medium mobility in one or other of the solvents used, over-run chromatography might offer a method for better separating them. In over-run chromatography, instead of the chromatogram being withdrawn from the solvent at the end of a normal run, the solvent is permitted to over-run, i.e. to drip off the lower edge of the paper. (It is worth cutting serrations into the lower edge to ensure that the solvent drips off evenly.) The degree of over-running can be controlled by the amount of solvent made available. Thus, a normal run would use 50 ml per chromatogram and this can be scaled up for double or triple over-runs by supplying 100 to 150 ml per chromatogram. It can be assumed that doubling the amount of solvent will result in the spots migrating double the distance, and so inspection of the original poorly resolved p.c. will give a good guide as to the degree of over-running possible, before the spots of interest themselves drip off.

An alternative procedure is to use a different solvent for the chromatography. Some of the many available are listed in Table 2.2 together with comments on their best applications. Most of these should be tried only after TBA or BAW which are usually the best in terms of solvent power and spot resolution.

Table 2.2
Useful solvents for the paper chromatographic analysis of flavonoids

Solvent	Solvent composition (approximate running time)[b]	Recommended uses[a]
BAW	*n*-BuOH:HOAc:H_2O (4:1:5) mixed thoroughly in a separating funnel, upper phase used (17 h)	*Glycosides*, *aglycones* and *sugars* (24–48 h), glucose and galactose not separated
TBA	*t*-BuOH:HOAc:H_2O (3:1:1) (24 h)	As for BAW
CAW	$CHCl_3$:HOAc:H_2O (30:15:2) mixed thoroughly in a separating funnel, excess water discarded (5–7 h)	*Aglycones* (especially methylated), *acylated glycosides*, *methylated glycosides*. Good for separation of isorhamnetin-kaempferol/syringetin/laricitrin type mixtures (same B-ring -OH pattern)
Forestal	HOAc:H_2O:HCl (30:10:3) (15 h)	*Flavone*, *flavonol* and *anthocyanidin aglycones*. Often will not separate glycosides from aglycones
"Formic acid"	HCOOH:H_2O:HCl (5:3:2)	*Anthocyanidins*. Will separate glycosides from aglycones (cf. forestal)
Bu/HCl	*n*-BuOH:2M HCl (1:1) mixed thoroughly in a separating funnel and equilibrated overnight (24 h)	*Anthocyanins* (chromatogram should be equilibrated in solvent vapour before running)
BEW	*n*-BuOH:EtOH:H_2O (4:1:2.2) if immiscible at room temperature add a little more EtOH (20–24 h)	Low capacity solvent, ideal for final clean-up of near pure *Glycosides*, *Sugars*
Benzene–acetic acid–water	Bz:HOAc:H_2O (125:72:3) solvent is volatile, thus good tank saturation is necessary (4 h)	*Aglycones*, phenolic acids
EPAW	EtOAc:Pyr:HOAc:H_2O (36:36:7:21	*Sugars*, especially t.l.c.

Solvent	Solvent composition (approximate running time)	Recommended uses[a]
Phenol	Phenol (4 g):H_2O(1 ml). Dry chromatograms in a vented oven (50–60°C) for 2–3 h followed by 24 h in a fume cupboard (to remove phenol) (24 h)	*Glycosides*. Separates vicenin-2 and schaftoside; glucuronides < glycosides; rhamnosides > glucosides. Separates glucose from both allose and galactose. *Sugars* (30–48 h)
H_2O	(4 h)	*Glycosides*. Generally *O*-glucuronides > *C*-glucosides > *O*-glucosides. Ideal for distinguishing *O*-glucuronides from other *O*-glycosides
HOAc	5% HOAc (3 h)	*Polyglycosides*
	15% HOAc (5 h)	*Glycosides*. Good for distinguishing mono-, di- and tri-glycosides
	50% HOAc (12 h) best "over-run"	*Aglycones*
1% HCl	(5 h)	*Anthocyanins*, phenolic acids
BBPW	*n*-BuOH:Bz:Pyr:H_2O (5:1:3:3) (20 h)	*Sugars*. Separates glucose and galactose and xylose and arabinose

[a] The symbols <, > refer to spot mobilities.
[b] For large sheets.

Flavonoid workers are divided over their preference for either BAW or TBA and it would seem appropriate here to mention the "pros" and "cons" of each. The advantage of BAW over TBA is the shorter development time required per p.c. run (16–18 h at 15°C compared with 24–36 h for TBA). This, however, is offset by the relative disadvantages which are (a) the lower mobility (*c.* 20%) of most flavonoids in BAW (e.g. Pangon *et al.*, 1974), and (b) the necessity to equilibrate and separate two phases during its preparation (TBA is a completely miscible solvent mixture). The compositions of both solvents change with time due to internal esterification and so neither should be stored for more than a few days before use.

2.3.4 R_f values

The mobility of any one compound in a particular solvent is referred to as its R_f value in that solvent, and in theory should be a reproducible characteristic of the compound. The R_f value is defined as the distance travelled by the compound (from the origin) divided by the distance travelled by the solvent front (from the origin). Hence, R_f values are always less than 1·0. A comparison of the R_f values of an unknown flavonoid with those published for known flavonoids of the same type is a useful method for comparing the unknown with flavonoids not available in the laboratory. R_f data for flavonoid standards are available from a number of sources including the sizeable compilations of Jiracek and Prochazka (1963) and Mabry *et al.* (1970) (TBA, 15% HOAc). It should be stressed, however, that R_f values are only approximately reproducible from laboratory to laboratory and, for this reason, apparent correlations with literature values must be taken only as a guide to identification pending side by side chromatography of the two compounds on the same p.c. (See Chapter 7 on the confirmation of identity.)

2.3.5 *Spray reagents and high-sensitivity detection*

By spraying the completed chromatograms with different reagents, a limited amount of flavonoid structure information can be gained. However, this is generally better obtained in other ways. In the author's experience the most advantageous application of spray reagents is that of heightening the sensitivity of flavonoid spot detection, e.g. in weakly loaded paper chromatograms or when searching for invisible or minor constituents. Four particularly useful sprays for this purpose are the following:

i. *$AlCl_3$* The 5% $AlCl_3$ solution used for u.v.-visible spectroscopy (see Chapter 3) when sprayed on to a p.c. which is subsequently dried, reveals all 5-hydroxy-flavonoids as fluorescent yellow spots when viewed with u.v. light (366 nm). Otherwise invisible spots are often revealed.

ii. *Diphenyl-boric acid-ethanolamine complex (Naturstoffreagenz A)* Application of a 1% solution in MeOH to the p.c. reveals (after drying) all 3′,4′-dihydroxy-flavones and -flavonols as orange spots (u.v. or visible). 4′-Hydroxy equivalents appear yellow-green.

iii. *Diazotized sulphanilic acid* The reagent is prepared as follows. A 0·3% solution of sulphanilic acid in 8% HCl (25 ml) is mixed with 5% sodium nitrite solution (1·5 ml) just prior to use. The p.c. is sprayed with this and then with a 20% solution of sodium carbonate before drying. Most compounds with free phenolic hydroxyl groups will show as yellow, orange or red spots.

iv. *Vanillin-HCl* 5% Vanillin in EtOH is mixed with concentrated HCl in the ratio 4:1 just prior to use. Red or purplish-red spots are produced by catechins and proanthocyanidins immediately on spraying and warming (hairdryer), and by flavanones and dihydroflavonols more slowly. The reagent reacts with all flavonoids possessing a phloroglucinol A-ring oxidation pattern in combination with a saturated C-ring. Such compounds are often not visible on paper chromatograms in u.v. light.

2.3.6 *Preparative p.c. for the isolation of flavonoids*

In order to isolate individual flavonoids for further study it is necessary to scale up the 2D-p.c. procedure if p.c. is to be used for this purpose. Loading the 2D paper chromatograms to maximum capacity and running a large number of them is one method of separating quantities of each flavonoid. The equivalent spots from each p.c. are then cut out, combined, cut into small strips and slurried with a suitable solvent (usually MeOH for aglycones and MeOH:H_2O, 1:1 for glycosides). After several hours of contact with occasional (or preferrably continual) shaking, the liquid is decanted. For efficient extraction this procedure should be carried out two or three times. The extracts are then combined, filtered and evaporated to dryness. Further purification if necessary may then be carried out by p.c. in a different solvent.

A more effective method, if applicable, is one-dimensional p.c. (1D-p.c.). If the 2D-p.c. indicates that separation of the components of interest could be carried out predominantly by using only one of the solvents, then 1D-p.c. is an appropriate method. Instead of applying the extract as a circular spot at the origin, the extract is streaked right across the paper to produce a 1–3-cm wide band at the origin (see Fig. 2.1(c)). This technique permits the chromatographic separation of 10–15 times the amount of extract separated on a 2D-p.c. and thus minimizes both the time and expense involved. After development of the 1D-p.c., the resultant bands may either be cut into small portions and extracted by slurrying with solvent (see above), or better, cut out as bands and eluted in a chromatography cabinet into a beaker (see Fig. 2.1(d)). This latter method, although slower, gives a concentrated, fibre-free solution of flavonoid. If necessary, the 1D-p.c. separation may be used in conjunction with other 1D- or 2D-p.c. runs to produce pure flavonoids. Potential also exists for further scaling up through the use of thicker chromatographic paper, e.g. Whatman No 17 (which has almost 3 times the capacity of 3MM but lower resolution).

Flavonoids isolated by p.c. may still contain some soluble polysaccharide impurities and are best further purified by sephadex LH-20 chromatography (see section 2.4.3).

2.3.7 *Hints on solvent evaporation*

A labour-intensive step in the isolation of the many pure flavonoids often produced by the above procedures is the evaporation of the solutions produced. Heat-assisted vacuum rotary evaporation is the obvious method of choice for rapid evaporation of aqueous solutions. Most flavonoids and their glycosides are stable under these conditions, but if decomposition is suspected, freeze-drying of the solutions is a suitable alternative. Ultimately, all large volumes are reduced to small volumes and each must be transferred to a separate vial and evaporated. This final evaporation may be conveniently carried out under vacuum in a vacuum desiccator. Large numbers of small volume samples can be evaporated simultaneously, and by using freshly dried silica-gel (containing indicator) and water-pump vacuum, most 1–2 ml solutions will evaporate completely overnight. If a more rapid evaporation of one sample is desired, the vial may be placed in a small round-bottomed flask and the flask placed on a rotary evaporator. Yet another technique for this purpose is to heat the vial on a water-bath whilst blowing a stream of nitrogen into the top. This results in very rapid evaporation.

2.4 Column chromatography—for large-scale separations

Using this technique the isolation of flavonoids can be scaled up almost to industrial scale. The technique basically involves application of a flavonoid mixture (in solution) to a column of powdered adsorbent (such as cellulose, silica or polyamide) and subsequent sequential elution of individual components with appropriate solvents. The column is simply a glass tube fitted with a tap at one end, with dimensions such that the diameter to length ratio is in the range 1:10 to 1:30 (see Fig. 2.3). The size (volume) required for any particular separation can be roughly calculated once the weight of the flavonoid mixture is known. It is generally considered that for separations based on partitioning (i.e. most cellulose and silica chromatography) the sample to column packing ratio should be in the range 1:50 to 1:500, the latter ratio being more appropriate to complex mixtures and the former to simple mixtures. With polyamide (and to some extent silica) columns, sample loadings in the 1:50 range are generally satisfactory.

Column packings should be chosen that are marketed specifically for column chromatography as the particle size is important. If it is too small, the flow rate of the eluting solvent may be unacceptably slow and if too large the chromatographic separation of components may be poor. Commercially available packings are usually in the 100–300 mesh range.

2.4.1 *Techniques of column packing and sample application*

Packing of the column should be carried out with care, the objective being to produce a homogeneously packed column. If the column does not have a built-in sintered glass filter in the neck, it will be necessary first to plug the neck of the column with a wad of glass or cotton wool. This should then be covered to a height of about 10 cm with the eluting solvent (see below). The column packing is then slurried in a beaker with the same solvent and poured carefully into the column, preferably all in one continuous process to avoid layering. The packing is then permitted to settle and the excess solvent drained off. With polyamide (and to some extent cellulose too) it is advisable to presoak the packing for up to an hour to permit swelling to take place before loading the column. This procedure has the added advantage of removing soluble material from the packing if the solvent is renewed just prior to filling the column. Wash the column thoroughly when packed.

The first step in column chromatography is to apply the sample solution to the top of the column in such a way that a narrow band is formed for further elution. To this end the sample should be dissolved in a minimum volume of solvent. The solvent used should be one of those selected for the later elution (see below) and preferably the one of lowest polarity, although

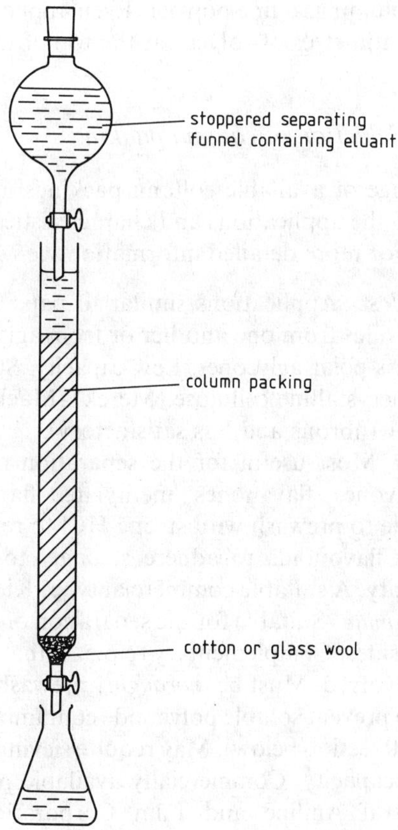

Fig. 2.3. Apparatus for column chromatography.

this is not always possible due to solubility difficulties. Application of the concentrated solution to the top of the column should be carried out with care to avoid disturbing the surface and use of a pipette is recommended for this purpose. The sample concentrate is then permitted to seep slowly into the column by opening the column tap slightly. Once the top of the packing is exposed again, the tap is closed, and some of the eluting solvent cautiously applied (again by pipette). This too is allowed to seep into the column before the bulk of the eluting solvent is added and the chromatographic process begun.

If solubility problems prevent the application of the sample in the above manner, the technique of pre-adsorption may be used. The sample is dissolved in a small amount of any suitable solvent and mixed in a beaker with a small amount of adsorbent (enough to form a 1–3-cm band on top of the prepared column). The adsorbent is then dried and applied evenly to the top of the column as a fine powder. Even application is aided by the presence of a few centimetres of solvent at the top of the column.

2.4.2 *Selection of column packing*

The range of available column packings is vast and the list below gives a guide to the applications and characteristics of a number of the more useful types. For more detailed information see Markham (1975).

Cellulose Applications similar to paper, i.e. ideal for the separation of glycosides from one another or from aglycones, and for the separation of the less-polar aglycones. Low capacity. Suitable commercial types include: microcrystalline cellulose (Merck) (Macheray and Nagel) and Whatman CF-11 (fibrous and less satisfactory).

Silica Most useful for the separation of the less polar aglycones, e.g. isoflavones, flavanones, methylated flavones and flavonols. Often advisable to prewash with strong HCl to remove traces of iron which cause many flavonoids to adhere strongly to the column packing. Medium capacity. A suitable commercial type: Kieselgel 60, 70–230 mesh (Merck).

Polyamide Suitable for the separation of all flavonoids, although ideal for glycosides. Complementary to p.c. in that different solvents and adsorbent are involved. Must be *thoroughly* prewashed with MeOH and H_2O before use to prevent soluble polyamide contaminating all fractions (see Sephadex LH-20 section below). May require sieving to obtain adequate particle size. High capacity. Commercially available polyamides include: Polyclar AT (General Aniline and Film Corporation), Polypenco 66D (Polymer Corporation) and Polyamide (Woelm).

Sephadex gel (G-series) Designed to separate mixtures predominantly on the basis of molecular size (when aqueous solvents are used), the larger molecules being eluted first. In this mode sephadex is useful for the separation of polyglycosides of different molecular weight. When organic solvents are used for elution, G-series sephadex gels behave chromatographically like cellulose but with higher capacity due to regular particle size. In either mode the gel must be swollen by presoaking in the eluant for 12 h before column packing. Commercially available products include: G-10 (for mol. wt 0–700), and G-25 (mol. wt 100–1500).

Sephadex gel (LH-20) Pharmacia sephadex LH-20 is designed specifically for use with organic solvents, and operates in both modes. Since it produces residue-free eluant, LH-20 is ideally suited to final clean-up of flavonoid aglycones and glycosides which have been isolated from paper, cellulose, silica or polyamide. MeOH is generally a suitable solvent, although some water may be needed initially to dissolve the flavonoid. Presoaking of the gel (as above) in MeOH is necessary.

2.4.3 *Solvent selection and column elution*

To a large extent, the solvent most suited to a particular separation must be determined experimentally and it is often simplest to explore possibilities by thin-layer chromatography (q.v.) prior to running a column. In general, however, the column should be run with a sequence of solvents or solvent mixtures beginning with the least polar and working up gradually to the most polar. The relative polarities of commonly used solvents are indicated by the eluotropic series listed below.

Cellulose, silica (Listed in order of decreasing solvent polarity) H_2O, formic acid, MeOH, HOAc, isopropyl alcohol (*i*-PrOH), acetone, *n*-PrOH, tetrahydrofuran, *t*-BuOH, 2°-BuOH, methyl ethyl ketone, *n*-amyl alcohol, ethyl acetate, ether, $CHCl_3$, benzene. The relationship between the polarities of many of these solvents is illustrated in Fig. 2.4 in which any vertical line indicates different solvent mixtures of equivalent eluting power, e.g. the dotted line indicates that 50:50 hydrocarbon:ethyl acetate has the same eluting power as 33:67 hydrocarbon:acetone and 8:92 ether:ethyl acetate.

Polyamide (Listed in order of decreasing effectiveness as eluant) aq. urea, dimethyl formamide, formamide, aq. NaOH, acetone, MeOH, H_2O, dichloroethane or $CHCl_3$.

Recommended "opening gambits" in the selection of solvent systems from each of these lists are as follows, the first mentioned solvent being the starting solvent which is gradually changed over to the second and so on.

Cellulose (general)	5% aq. MeOH→H_2O
(glycosides)	H_2O→15% HOAc (v/v) *or* 2°-BuOH:HOAc:H_2O (70:5:25). Choice depends on p.c. data, the second solvent approximating to TBA or BAW but is one phase and readily evaporated
Silica (less polar aglycones)	benzene: $CHCl_3$ (1:1)→$CHCl_3$→EtOAc
Polyamide (glycosides)	H_2O→MeOH *or* $CHCl_3$:MeOH→MeOH
(aglycones)	$CHCl_3$:EtOH: methyl ethyl ketone:acetone (40:20:5:1)→decreasing level of $CHCl_3$
Sephadex G (glycosides)	H_2O (for mol. wt separation)
LH (aglycones)	MeOH (increasing polarity with H_2O or decreasing polarity with *n*-BuOH)
LH (glycosides)	MeOH→30% aq. MeOH

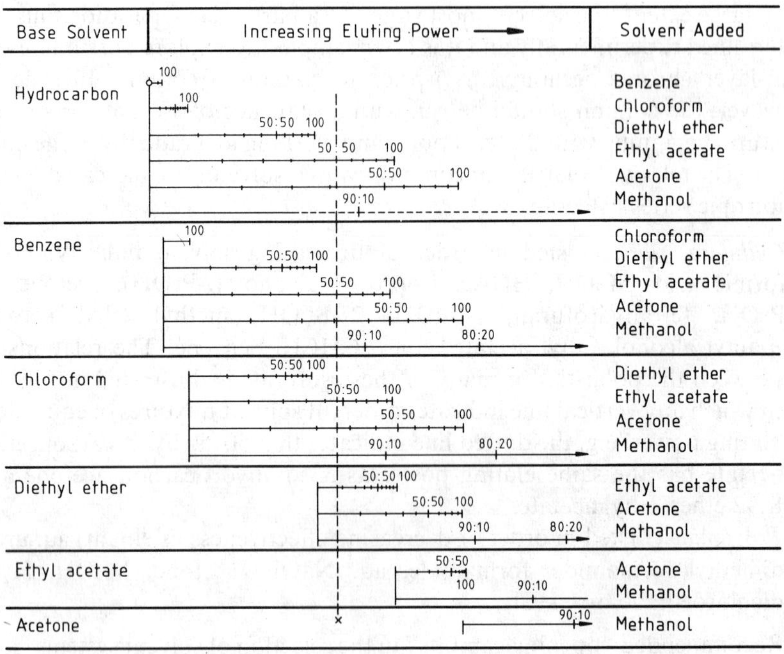

Fig. 2.4. Comparative polarities of some common solvents.

Modifications to, and extensions of, the solvent systems recommended on page 30, can be devised by reference to the eluotropic series listed above.

Elution of the flavonoids from the column must be carried out slowly if good resolution is required. The separating bands on the column may be visibly yellow or visible in u.v. light (366 nm), in which case collection of each band in a separate container is a simple procedure. If the bands are not visible, however, it is necessary to collect fractions at regular intervals (e.g. every 25 or 50 ml) and then to analyse each fraction by p.c. or t.l.c. to determine which fractions can be combined. An automated fraction collector is ideal for this purpose. The tedium of supplying solvent to the top of the column manually can be minimized by using a separating funnel as reservoir for automatic top-up (see Fig. 2.3).

2.5 Other useful chromatographic techniques

2.5.1 *Thin-layer chromatography (t.l.c.) for rapid microanalysis*

The main value of t.l.c. in flavonoid work is as a rapid analytical method which requires very small amounts of material. Much has been written about the technique itself (e.g. Kirchner, 1967; Stahl, 1969; Touchstone and Dobbins, 1978) and about its usefulness in flavonoid analysis (Stahl, 1969; Markham, 1975). For this reason discussion here will be kept to a minimum.

In the author's experience, t.l.c. is particularly useful for the following purposes:

a. investigating solvents for possible use in column chromatography;
b. analysis of fractions from column chromatography;
c. following the course of a reaction such as hydrolysis or methylation;
d. identification of flavonoids by cochromatography;
e. small scale isolation of pure flavonoids.

The absorbents and solvents used are generally similar to those discussed in detail for p.c. and column chromatography (but see below) and methods of spot detection are largely as outlined for p.c. Plastic-backed t.l.c. plates are especially recommended as they are transparent to u.v. light (for spot detection) and can readily be cut to any size. Such plates are marketed by Schleicher and Schüll (cellulose F1440; kieselgel F1500; micropolyamide A1700), Merck (cellulose; kieselgel; polyamide 11F) and Macherey-Nagel (polygram cellulose 300 u.v.). These 20×20 cm sheets are conveniently cut into 8×4 cm plates and chromatography can be carried out in a covered beaker, or better, in standard $9 \times 9 \times 4$ cm glass tanks (marketed for example by Kodak and Desaga). Plates of this size normally develop in 10–100 minutes depending upon the solvent used.

Cellulose t.l.c. plates may be prepared using standard t.l.c. spreaders (e.g. Camag, Corning etc). Merck microcrystalline "Avicel" cellulose is especially suitable for this purpose and is best spread to a thickness of 0·25 mm (to avoid cracking of the layer). To spread five 20 × 20 cm plates, 20 g of cellulose (stirred vigourously with 70 ml H_2O for 5 min) will suffice.

Recommended solvent/adsorbent combinations for particular flavonoid types are listed below (refer to Table 2.2 for solvent symbols). Other systems have been discussed by Markham (1975).

Flavonoid glycosides	*Cellulose*—TBA, BAW. 5–40% HOAc, H_2O (for distinction of glucuronides—see section 4.1.5) *Polyamide*—H_2O: MeOH: MEK: acetylacetone, 13:3:3:1 (for glucuronides, MeOH:HOAc:H_2O, 18:1:1) *Silica*—EtOAc: Pyr: H_2O: MeOH, 80:20:10:5 (especially flavone *C*-glycosides)
Polar flavonoid aglycones (e.g. flavones, flavonols)	*Cellulose*—TBA, 50% HOAc, CAW, benzene: HOAc: H_2O (125:72:5) *Polyamide*—MeOH:HOAc:H_2O (18:1:1) *Silica*—benzene: Pyr: HCO_2H, 36:9:5 (especially biflavonyls)
Non-polar flavonoid aglycones (e.g. dihydroflavonoids, isoflavones, methylated flavones)	*Cellulose*—10–30% HOAc *Silica*—$CHCl_3$:MeOH (15:1 to 3:1)

2.5.2 *Paper electrophoresis for analysis of flavonoid bisulphates and glucuronides*

Paper electrophoresis is a technique of limited application in flavonoid analysis since, to be mobile, a flavonoid must be in an ionized state at the pH of the electrolyte. Examples of its useful application are in the recognition and identification of flavonoid bisulphates and in the distinction of glucuronides from glucosides. Thus at pH 2·2 (formate-acetate buffer) with Whatman No 3 paper and a potential of 400 V cm^{-1}, flavonoid bisulphates migrate towards the anode whereas other flavonoids do not (Kruezaler and Hahlbrock, 1973; Harborne, 1977). The distinction of glucuronides from glucosides at low levels is conveniently accomplished at pH 4 (0·01 M potassium hydrogen phthalate) on cellulose acetate sheet at 400 V (3 mA). Under these conditions the glucuronides migrate towards the anode but the glucosides do not (Markham, 1980). Shandon Southern electrophoresis units fitted with voKam power supplies are commonly used for this work. Some relative mobilities of different flavonoid bisulphates are listed by Harborne (1977).

2.5.3 *High performance liquid chromatography (h.p.l.c.) for quantitative microanalysis*

It is not the intention here to discuss the technique of h.p.l.c. which is adequately covered in the literature, e.g. Pryde and Gilbert (1979). However, in view of the rapidly increasing popularity of the technique, brief comment on its applicability to flavonoid analysis is warranted.

High performance liquid chromatography is basically a form of column chromatography which utilizes a column of packing material of small particle size and regular shape. Because of the fineness of the packing, pressures of up to 5000 lb in^{-2} are employed to achieve acceptable flow rates. The technique offers the researcher a method of *quantitatively* analysing the flavonoid components of a mixture at a high level of resolution and sensitivity (<50 ng) and it is the quantitative aspect of the analysis in particular which sets it apart from other chromatographic methods. Quantification is achieved by automatically monitoring the eluant leaving the column by means of a variable wavelength u.v. spectromonitor and the "chromatogram" is traced out as a series of peaks on a chart.

A wide range of packing/solvent combinations have been reported in the literature for flavonoids and a useful review of these has been published by Kingston (1979). It is clear that for most applications reversed phase columns (in which a hydrocarbon is bonded to the silica packing) of the μ-Bondapak C-18 (Waters Associates) type are suitable. Solvents such as H_2O/MeOH, H_2O/MeOH/HOAc and H_2O/acetonitrile (in varying proportions) have been used successfully to chromatograph flavones, flavonols, dihydroflavonoids, catechins, anthocyanidins and flavonoid glycosides, and in some cases a changing solvent composition (solvent programming) has proven useful.

Disadvantages of the method include the expense of the pump/detector/recorder/column (minimum package), and the requirement to inject only particle-free solutions on to the column to avoid blockages and column deterioration. If sufficient of the flavonoid mixture is available, then quantitative analysis of the components is conveniently and cheaply accomplished by a combination of p.c. (to separate components) and u.v.-visible spectroscopy (to quantify components).

2.6 Recrystallization techniques

If sufficient flavonoid is available (i.e. 4 mg or more) a most effective method of purification, both from other minor flavonoids and from soluble chromatographic material, is recrystallization. The melting point of flavonoids purified in this manner is a useful characteristic for comparison of the unknown with previously isolated flavonoids, and a "mixed melting point" determination (q.v.) is an excellent method for establishing the identity of one flavonoid with another. Crystalline flavonoids are also the ideal starting material for sugar analysis since they are free from the soluble polysaccharides which invariably contaminate flavonoids isolated by p.c.

Method (i) By trial and error a solvent is found in which the flavonoid has limited solubility. H_2O or MeOH are often useful solvents for flavonoid glycosides, flavones and flavonols. A minimum volume of this solvent is used to dissolve the flavonoid completely at 100°C.

The solution is then filtered either by using an apparatus such as in Fig. 2.5, or, if the volume is small, by sucking it through a small ball of cotton wool with a Pasteur pipette. When left to cool (or evaporate) in a small conical flask, the solution should slowly yield crystals. Refrigeration may help.

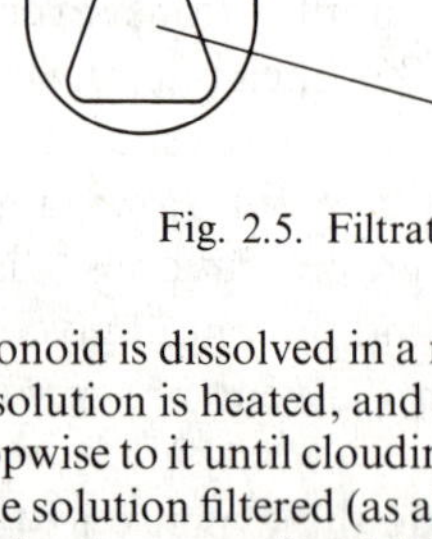

Fig. 2.5. Filtration apparatus.

Method (ii) The flavonoid is dissolved in a minimum volume of a solvent in which it is readily soluble. This solution is heated, and a miscible solvent in which the flavonoid is insoluble is added dropwise to it until cloudiness appears. A few drops of the first solvent are then added and the solution filtered (as above) and left to cool and produce crystals. Suitable solvent combinations are often MeOH/H_2O for flavonoid glycosides, flavones and flavonols and diethyl ether/petroleum ether (60–80°C b.p.) for less polar aglycones. It is advantageous to select solvent combinations such that the most volatile solvent is that in which the flavonoid is most soluble. Thus in time the flavonoid must come out of solution as the level of this solvent is reduced by evaporation.

Method (iii) By the use of a Bolton extractor (Fig. 2.6), flavonoids which prove to be insufficiently soluble in the normal range of solvents may be recrystallized. The flavonoid in the soxhlet thimble is gradually dissolved in the refluxing solvent and concentrates in the round bottom flask below. When most of the flavonoid has been dissolved, the solution is permitted to cool and the sparingly soluble flavonoid should crystallize out. If it fails to do so, some of the solvent may be removed by distillation to speed up the process.

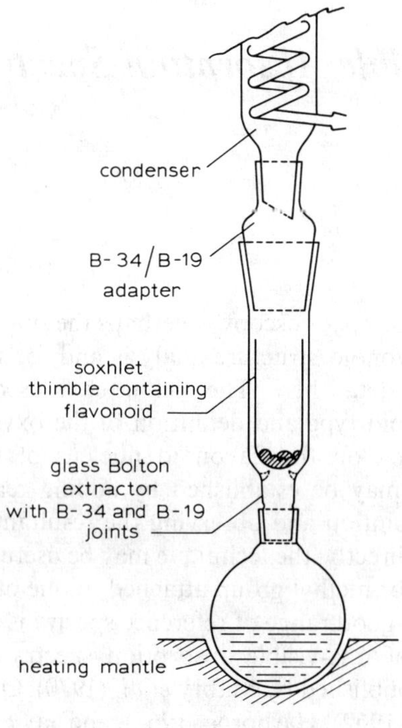

Fig. 2.6. Bolton extractor in use.

3

Ultraviolet-visible Absorption Spectroscopy

3.1 Introduction

Ultraviolet-visible absorption spectroscopy is perhaps the single most useful technique available for flavonoid structure analysis and for that reason it will be dealt with in some detail here. The technique is used to aid both identification of the flavonoid type and definition of the oxygenation pattern. In addition, the siting on the flavonoid nucleus of unsubstituted phenolic hydroxyl groups may be established by adding reagents ("shift reagents") to the sample solution and observing the resultant shifts in the absorption peaks. Thus, indirectly, the technique may be useful in determining the location of a sugar or methyl group attached to one of the phenolic hydroxyls. Availability of a good range of reference spectra is an invaluable aid in the interpretation of u.v.-visible absorption spectra and a useful general selection has been published by Mabry *et al.* (1970). Other compilations include those of Jurd (1962), Harborne (1967), and Jay *et al.* (1975).

A major advantage of this method is the very small amount of flavonoid required for full analysis (usually *c.* 0·1 mg). In practice, few workers bother to weigh accurately the sample used, preferring instead to make up a stock solution of about 0·1 mg per 10 ml in AR methanol and then dilute it until the absorption level (absorbance) of the major peak is about 0·6. This stock solution is then used for all subsequent measurements.

For quantitative analysis of a known flavonoid the relationship known as the Beer–Lambert law is used:

$$\text{Beer–Lambert Equation:} \quad A = \varepsilon \times c \times d.$$

In this equation A = absorbance (read from the instrument), ε = extinction coefficient, c = concentration of flavonoid in moles (gram/molecular weight) per litre, and d = path length of cell (cm).

Quantification of a flavonoid in solution may thus be carried out if the extinction coefficient is known. A good compilation of extinction coefficients has been reported by Jurd (1962), but if the one required is not available it may be calculated by measuring the absorbance of a known concentration of the studied flavonoid (obtained from another source).

3.2 The flavonoid spectrum—general

The flavonoid spectrum is usually determined for a methanol (MeOH, AR or equivalent) or, less satisfactorily (Porter and Markham, 1970), ethanol (EtOH) solution of the flavonoid. However, 0.4 M methanolic HCl is required for anthocyanins. The spectrum typically consists of two absorption maxima in the ranges 240–285 nm (band II) and 300–550 (band I). The precise position and relative intensities of these maxima give valuable information on the nature of the flavonoid and its oxygenation pattern. Typical spectra of the major flavonoid types with equivalent (5,7,4′) oxygenation patterns are presented in Figure 3.1 below. Characteristic features of these spectra are the low relative intensities of Band I in dihydroflavones, dihydroflavonols, and isoflavones, and the long wavelength position of band I in the spectra of chalcones, aurones and anthocyanins. These characteristics are relatively invariant even with changing oxygenation pattern, although the ranges of absorption maxima for different flavonoid types do overlap as a result of variations in oxygenation patterns. Some guide as to the expected ranges of the principal maxima for each flavonoid type is given in Table 3.1.

Variation within these ranges will depend on the hydroxylation pattern and on the degree of substitution of the hydroxyls. The reader is referred especially to Mabry *et al.* (1970), or Harborne (1967) for anthocyanins, for graphically presented examples of the effects brought about by these changes. In brief they are the following:

1. Changes in the substitution of the A-ring tend to be reflected in the band II absorption while alterations in the substitution of the B- and C-rings tend to be more apparent from the band I absorption.
2. Additional oxygenation (especially hydroxylation) generally causes a shift of the appropriate band to longer wavelengths, e.g. band I in 3,5,7-triOH flavone, 359 nm; 3,5,7,4′-OH flavone, 367 nm; 3,5,7,3′,4′-OH flavone, 370 nm and 3,5,7,3′,4′,5′-OH flavone, 374 nm.
3. Methylation or glycosylation (especially of 3,5,7 and 4′ hydroxyls) causes band shifts to shorter wavelengths. The nature of the sugar in glycosides is normally of no consequence.

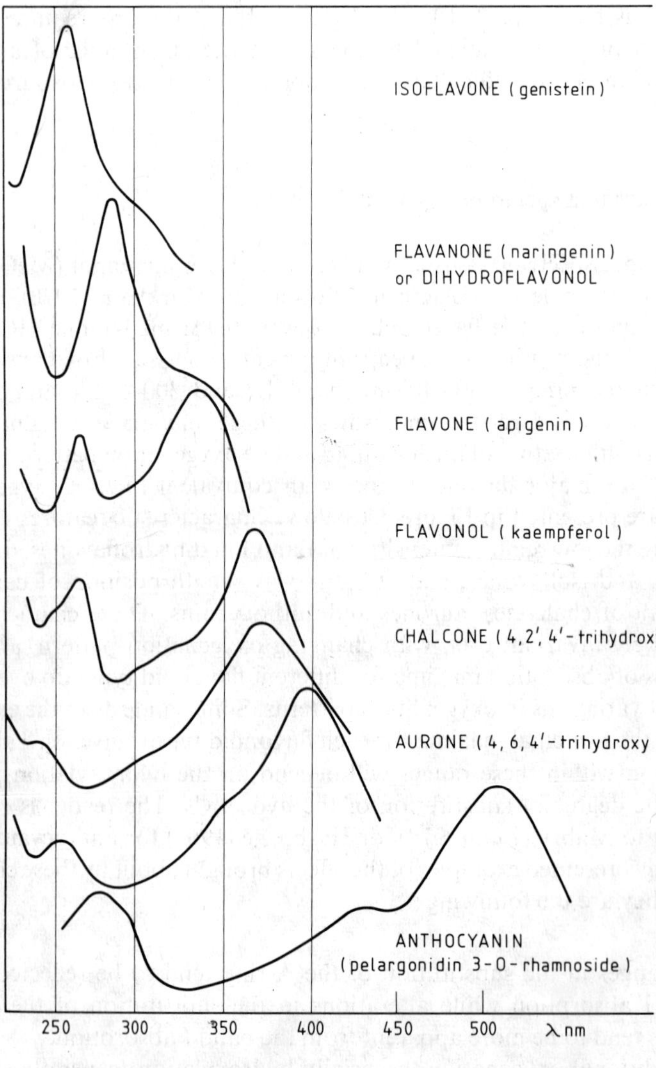

Fig. 3.1. Ultraviolet-visible absorption spectra of different flavonoid types with equivalent hydroxylation patterns.

4. Acetylation tends to nullify the effect of a phenolic hydroxyl group on the spectrum.
5. The presence of cinnamic acids as acyl functions on a flavonoid can be detected by the presence of an absorption band at *c*. 320 nm in flavonoids that themselves lack significant absorption in this region (e.g. anthocyanins, Harborne, 1967).
6. In flavones and flavonols the presence of a 3′,4′-diOH system is generally evidenced by a second peak (sometimes a shoulder) in band II.

Table 3.1
Ultraviolet-visible absorption ranges for flavonoids

Band II (nm)	Band I (nm)	Flavonoid type
250–280	310–350	Flavone
250–280	330–360	Flavonols (3-OH substituted)
250–280	350–385	Flavonols (3-OH free)
245–275	310–330 shoulder	Isoflavones
	c. 320 peak	Isoflavones (5-deoxy-6,7-dioxygenated)
275–295	300–330 shoulder	Flavanones and dihydroflavonols
230–270 (low intensity)	340–390	Chalcones
230–270 (low intensity)	380–430	Aurones
270–280	465–560	Anthocyanidins and anthocyanins

3.3 Additional information gained by use of shift reagents

3.3.1 *Preparation of shift reagents*

Sodium methoxide (NaOMe) Approximately 2·5 g of metallic sodium (best handled and weighed under petroleum ether or hexane in a beaker) is cut into small portions and added *cautiously* to 100 ml of MeOH (quality as for sample). The resultant solution should be stored in a plastic-stoppered glass bottle. 2 M aq. NaOH is a suitable alternative to NaOMe.

Sodium acetate (NaOAc) Powdered, anhydrous, AR NaOAc is used. Fusion of the NaOAc to remove residual HOAc (Mabry *et al*., 1970) is not normally carried out.

Aluminium chloride ($AlCl_3$) About 5 g of fresh, dry, $AlCl_3$ (which should fizz when added to H_2O!) is added cautiously to 100 ml of AR MeOH.

Residual insoluble material will normally dissolve with time. Store in a plastic stoppered bottle.

Hydrochloric acid (HCl) Concentrated reagent grade HCl (50 ml) is added to 100 ml of distilled H_2O.

Boric acid (H_3BO_3) Anhydrous, powdered, AR grade is used.

3.3.2 *Stepwise procedure for use of shift reagents*

a. After measurement of the spectrum of the sample in MeOH (the "MeOH" spectrum), 3 drops of NaOMe are added to the cuvette (1 cm, 2 ml vol.) and after mixing, the "NaOMe" spectrum is recorded. To check for sample decomposition the "NaOMe" spectrum is rerun after about 5 minutes. The sample is then discarded and the washed cell refilled with flavonoid stock solution.

b. Six drops of $AlCl_3$ reagent are added to the flavonoid solution, the sample mixed, and the "$AlCl_3$" spectrum measured. HCl (3 drops) is then added and, after mixing, the "$AlCl_3$/HCl" spectrum is measured. The sample is then discarded and the cell washed.

c. Powdered NaOAc is then added to fresh flavonoid stock solution in the cuvette in quantity such that about a 2 mm layer settles to the bottom. The mixture should be thoroughly shaken before the "NaOAc" spectrum is recorded. As in (a) above, a check for sample decomposition with time may be made at this stage. The "NaOAc/H_3BO_3" spectrum is then measured after addition of H_3BO_3 (quantity: about half as much as NaOAc) and mixing.

It is not necessary to add the shift reagents to the reference cell as well as to the sample, since the resultant improvement in the spectra is rarely worth the extra time involved. However, this procedure may be worthwhile if the amount of flavonoid available is small, since several of the shift reagents (NaOAc in particular) themselves absorb in the lower wavelength region and may lower resolution of band II in dilute sample solutions.

3.3.3 *Interpretation of the spectra*

Full interpretation of the set of spectra obtained from the above procedures can often only be achieved when information from other sources is also available. However, a good deal of information is obtainable from the spectra alone and this is discussed here.

The first step is to "type" the flavonoid using:

i. the general form of the "MeOH" spectrum (see Fig. 3.1);
ii. the wavelengths of the absorption bands (see Table 3.1);
iii. paper chromatographic data (see Chapter 2).

The second step is to consider the significance of changes induced in the spectrum by the various shift reagents. Interpretation of these changes is discussed below and tabulated keys based on flavonoid type are included for each shift reagent (abbreviations used in these Tables include: OH (hydroxyl), OR (substituted hydroxyl), + (bathochromic shift, i.e. to longer wavelength), *o*-diOH (*ortho*-related dihydroxyl system)).

The "NaOMe" spectrum The "NaOMe" spectrum represents that of the flavonoid with all phenolic hydroxyl groups ionized to some extent. It is therefore generally a good "fingerprint" indicator of the hydroxylation pattern as well as being useful for the detection of the more acidic hydroxyl groups in unsubstituted form. Degradation of the spectrum with time is a good indicator of the presence of alkali-sensitive groupings. Details of the interpretation are presented in Table 3.2 and a range of reference spectra is listed in Table 3.6 (see Fig. 3.3 for pictorial presentation).

The "NaOAc" spectrum Sodium acetate causes significant ionization of only the most acidic of the flavonoid hydroxyl groups (for relative ionization constants see Tyukavkina and Pogodaeva, 1975). Thus it is used primarily to detect the presence of a free 7-hydroxyl group (or equivalent). See Table 3.3 and Fig. 3.4.

The "NaOAc/H_3BO_3" spectrum NaOAc/H_3BO_3 bridges the two hydroxyls in an *ortho*-dihydroxy group and is used to detect their presence. See Table 3.4.

The "$AlCl_3$" and "$AlCl_3$/HCl" spectra By forming acid-stable complexes between hydroxyls and neighbouring ketones, and acid-labile complexes with *ortho*-dihydroxyl groups, these reagents can be used to detect both groupings. The "$AlCl_3$" spectrum thus represents the sum effect of *all* complexes on the spectrum, whilst the "$AlCl_3$/HCl" spectrum represents the effect only of the hydroxy-keto complexes (see Fig. 3.2, p. 45). It is worth noting that if ethanol is used as solvent (rather than methanol) the shifts obtained with $AlCl_3$ are very sensitive to traces of water (Porter and Markham, 1970). The interpretation of these spectra is summarized in Table 3.5. (See also Fig. 3.5 for pictorial presentation.)

Table 3.2
Interpretation of "NaOMe" spectra

Flavonoid type	Shift observed Band I	Band II	Interpretation guide
Flavones Flavonols	Continually reducing intensity (i.e. decomposition)		3,4′-OH, A-ring *o*-diOH, B-ring: 3 adjacent OH
	Stable +45 to 65 nm no decrease in intensity		4′-OH
	Stable +45 to 65 nm decreased intensity		3-OH, no free 4′-OH
	New band (cf. MeOH), 320–335 nm		7-OH
Isoflavones		No shift	No A-ring -OHs
Flavanones Dihydroflavonols		Decreasing intensity with time	A-ring *o*-diOH (slow decrease: B-ring *o*-diOH in isoflavones)
		Shift from *c.* 280 nm to *c.* 325 nm, increased intensity, but	5,7-OH flavanones and dihydroflavonols
		to 330–340 nm	7-OH, no free 5-OH
Chalcones Aurones	+80 to 95 nm (increased intensity)		4′-OH (aurones)
	+60 to 70 nm (increased intensity)		6-OH no 4′-oxygenation (aurones)
	Smaller shift		6-OH with 4′-oxygenation (aurones)
	+60 to 100 nm (increased intensity)		4-OH (chalcones)
	(No increase in intensity)		2-OH or 4′-OH and no 4-OH
	+40 to 50 nm		4′-OH (2′-OH or 4-OR also present)
Anthocyanidins Anthocyanins	All but 3-deoxyanthocyanidins decompose		Nil

Table 3.3
Interpretation of "NaOAc" spectra

Flavonoid type	Shift observed Band I	Band II	Interpretation guide
Flavones Flavonols Isoflavones		+5 to 20 nm (reduced if 6- or 8-oxygenation present)	7-OH
	Decreasing intensity with time		Alkali-sensitive groups, e.g. 6,7 or 7,8 or 3,4′-diOH
Flavanones Dihydroflavonols		+35 nm +60 nm	7-OH (with 5-OH) 7-OH (without 5-OH)
	Decreasing intensity with time		Alkali-sensitive groups e.g. 6,7 or 7,8-diOH
Chalcones Aurones	Bathochromic shift or long-wavelength shoulder		4′ and/or 4-OH (chalcones) 4′ and/or 6-OH (aurones)

Table 3.4
Interpretation of "NaOAc/H_3BO_3" spectra

Flavonoid type	Shift observed Band I	Band II	Interpretation guide
Flavones Flavonols Aurones Chalcones	+12 to 36 nm (relative to MeOH spectrum)		B-ring *o*-diOH
	Smaller shift		A-ring *o*-diOH (6,7 or 7,8)
Isoflavones Flavanones Dihydroflavonols		+10 to 15 nm (relative to MeOH spectrum)	A-ring *o*-diOH (6,7 or 7,8)

Table 3.5
Interpretation of "$AlCl_3$" and "$AlCl_3$/HCl" spectra

Flavonoid type (reagent)	Shift observed Band I	Band II	Interpretation guide
Flavones and flavonols ($AlCl_3$/HCl)	+35 to 55 nm		5-OH
	+17 to 20 nm		5-OH with 6-oxygenation
	No change		Possibly 5-OH with 6-prenyl group
	+50 to 60 nm		3-OH possible (with or without 5-OH)
($AlCl_3$)	$AlCl_3$/HCl shift plus 30 to 40 nm		B-ring *o*-diOH
	$AlCl_3$/HCl shift plus 20 to 25 nm		A-ring *o*-diOH (additive to B-ring *o*-diOH shift)
Isoflavones, flavanones and Dihydroflavonols ($AlCl_3$/HCl)		+10 to 14 nm	5-OH (isoflavones)
		+20 to 26 nm	5-OH (flavanones, dihydroflavonols)
($AlCl_3$)		$AlCl_3$/HCl shift, plus 11 to 30 nm	A-ring *o*-diOH (6,7 and 7,8)
		$AlCl_3$/HCl shift, plus 30 to 38 nm (NaOAc sensitive)	Dihydroflavonol with no 5-OH (additive to any *o*-diOH shift)
Aurones Chalcones ($AlCl_3$/HCl)	+48 to 64 nm		2′-OH (chalcones)
	+40 nm		2′-OH (chalcones) with 3′-oxygenation
	+60 to 70 nm		4-OH (aurones)
($AlCl_3$)	$AlCl_3$HCl shift, plus 40 to 70 nm		B-ring *o*-diOH
	Smaller increase		Possibly A-ring *o*-diOH

Table 3.5 (Cont.)

Flavonoid type (reagent)	Shift observed Band I	Band II	Interpretation guide
Anthocyanidins Anthocyanins ($AlCl_3$)	+25 to 35 nm (at pH 2–4)		o-diOH
	Larger shifts		Multiple o-diOH or o-diOH (3-deoxy-anthocyanidins)

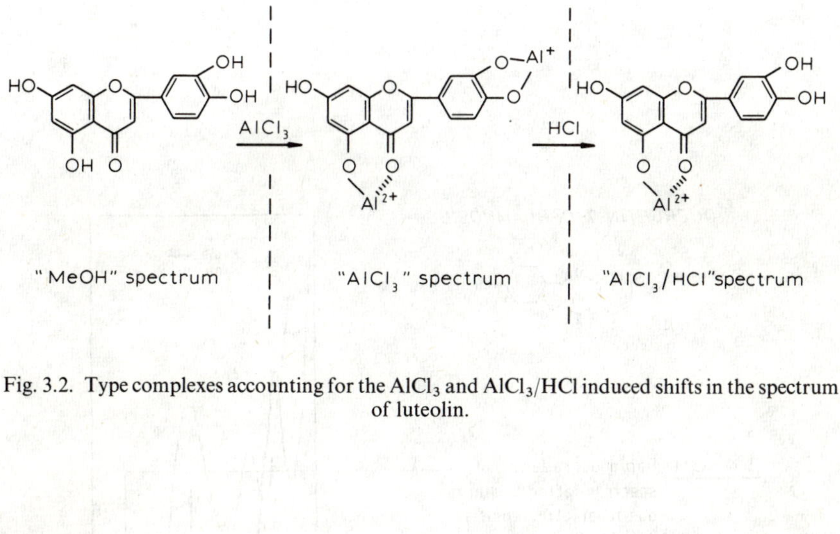

Fig. 3.2. Type complexes accounting for the $AlCl_3$ and $AlCl_3$/HCl induced shifts in the spectrum of luteolin.

3.3.4 *Spectra exemplifying shift reagent effects*

A selection of spectra is presented below which exemplify the identification of key flavonoid structure features by use of the NaOMe, NaOAc, $AlCl_3$ and $AlCl_3$/HCl shift reagents (Figs 3.3, 3.4 and 3.5)

Fig. 3.3 Examples of the usefulness of "NaOMe"
Fig. 3.4 An example of the usefulness of "NaOAc"
Fig. 3.5 An example of the usefulness of "$AlCl_3$" and "$AlCl_3$/HCl"

RUTIN

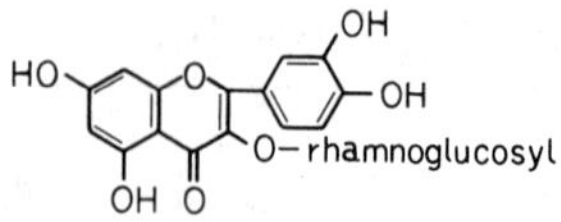

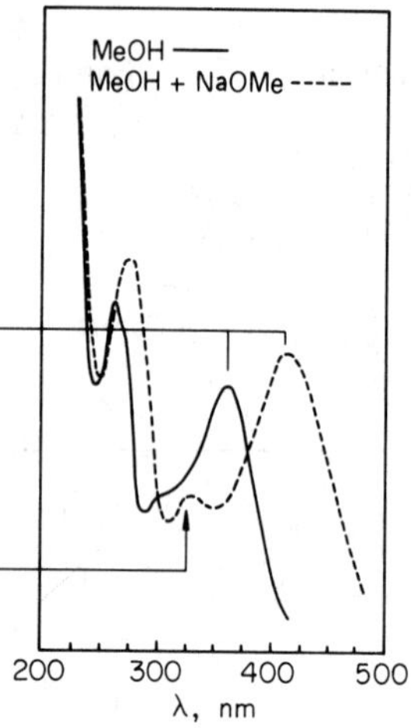

4′-OH 50 nm bathochromic shift and increasing intensity indicate 4′-OH

7-OH Free 7-OH indicated by new band at 327 nm

QUERCETIN 7-O-RHAMNOSIDE

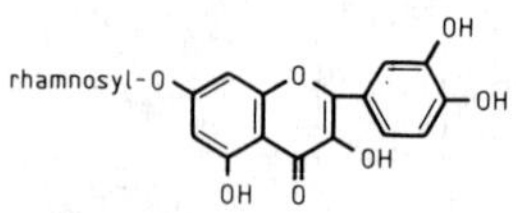

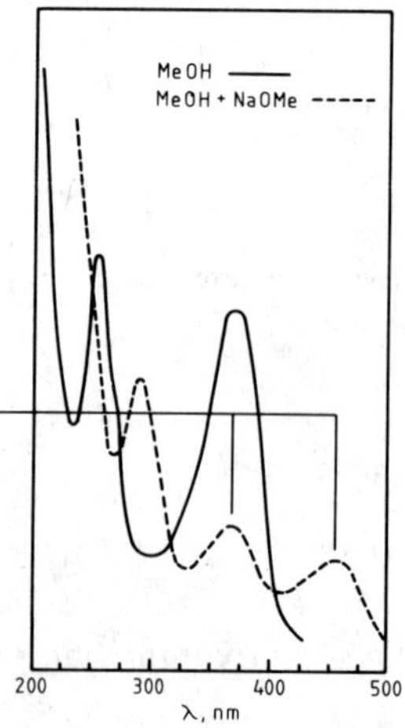

3,4′-OH Rapid degradation of spectrum after 5 minutes due to alkali-sensitive 3,3′,4′-OH system.

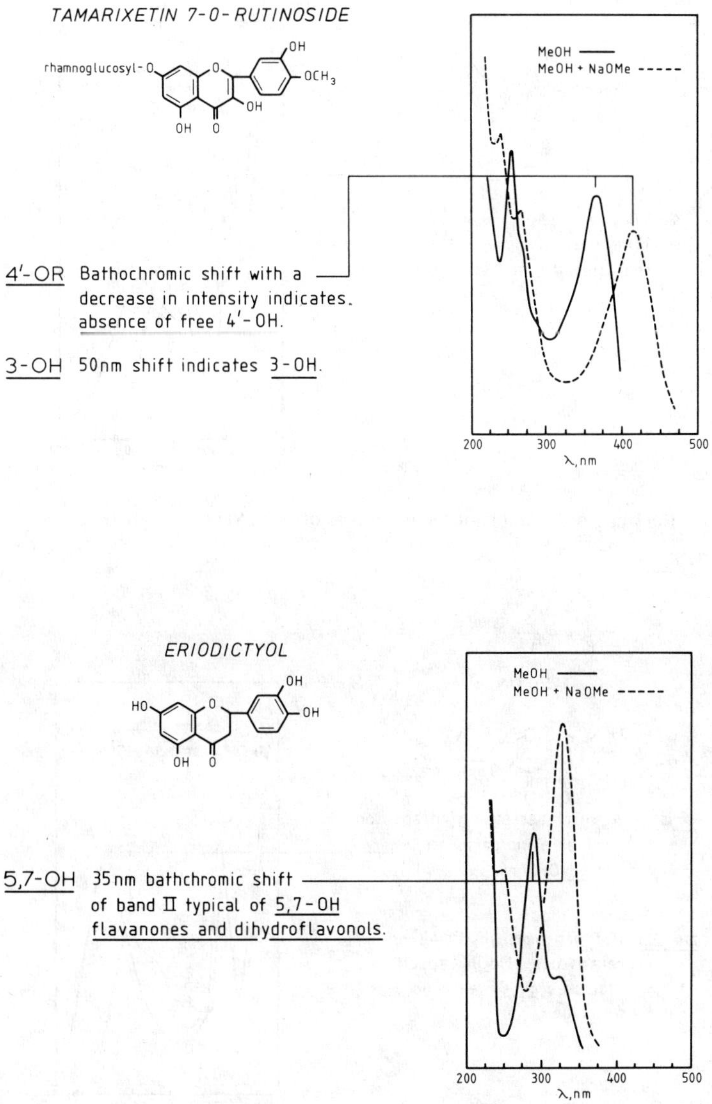

Fig. 3.3. Examples of the usefulness of the NaOMe shift reagent.

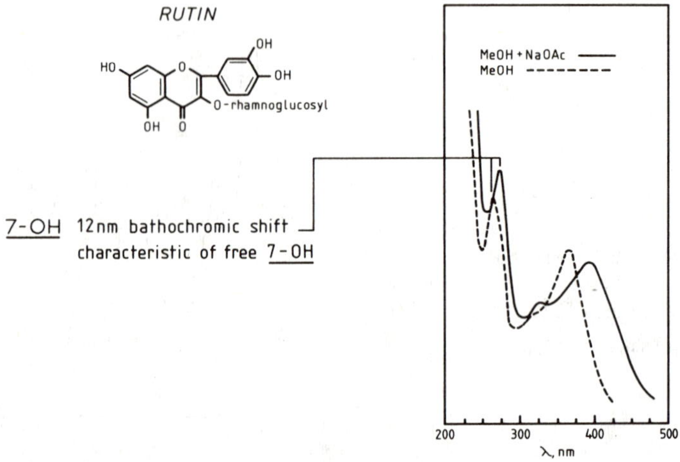

Fig. 3.4. An example of the usefulness of the "NaOAc" shift reagent.

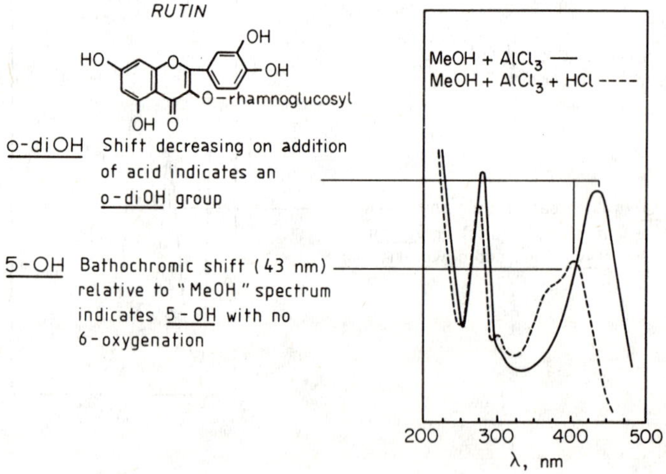

Fig. 3.5. An example of the usefulness of the $AlCl_3$ and $AlCl_3$/HCl shift reagents.

3.4 Sources of reference spectra

The following list is a guide to literature sources of reference spectra.

Most flavonoid types	Mabry *et al.* (1970)
	Harborne (1967)
	Jurd (1962)
	Litvinenko and Maksyutina (1965)
Flavones and Flavonols	Jay *et al.* (1975)
Anthocyanins	Harborne (1958)
	Nilsson (1973)
Aurones	Geissman and Harborne (1956)
	Huke *et al.* (1969)
Biflavones	Dossaji *et al.* (1975)
Chalcones	Jurd and Horowitz (1961)
Isoflavones	Horowitz and Jurd (1961)

A selection of the more commonly required spectra from some of these sources is presented in Table 3.6.

Table 3.6
Ultraviolet-visible absorption spectra of some commonly encountered flavonoids[a, b]

Flavones	MeOH (λ_{max}nm)	NaOMe (λ_{max}nm)
Chrysin (5,7)	247sh, 268, 313	288, 263sh, 277, 361
Baicalein (5,6,7)	247sh, 274, 323	257, 366, 410sh (dec)
Apigenin (5,7,4′)	267, 296sh, 336	275, 324, 392
Apigenin 7-*O*-glucoside	268, 333	245sh, 269, 301sh, 386
Apigenin 8-*C*-glucoside	270, 302sh, 336	279, 329, 395
Acacetin (5,7,4′)	269, 303sh, 327	276, 295sh, 364
Acacetin 7-*O*-glucoside	268, 324	244sh, 287, 357
Luteolin (5,7,3′,4′)	253, 267, 291sh, 349	266sh, 329sh, 401
Luteolin 7-*O*-glucoside	255, 267sh, 348	263, 300sh, 394
Luteolin 8-*C*-glucoside	255, 267, 293sh, 346	268, 278sh, 334sh, 405
Chrysoeriol (5,7,3′,4′)	241, 249sh, 269, 347	264, 275sh, 329sh, 405
Diosmetin (5,7,3′,4′)	252, 267, 291sh, 344	270, 303sh, 386
Tricin (5,7,3′,4′,5′)	244sh, 269, 299sh, 350	263, 275sh, 330, 416
Flavonols		
Galangin (3,5,7)	267, 305sh, 359	280, 327sh, 412
Galangin 3-methyl ether	266, 312sh, 340sh	276, 360
Kaempferol (3,5,7,4′)	266, 294sh, 322sh, 367	278, 316, 416 (dec)
Kaempferol 3-*O*-glucoside	264, 301sh, 350	273, 324, 398

Table 3.6 (Cont.)

	MeOH (λ_{max}nm)	NaOMe (λ_{max}nm)
Kaempferol 7-O-glucoside	264, 321, 362	270, 294sh, 358, 450 (slow dec)
Kaempferol 3,7-*O*-glucoside	265, 323sh, 350	272, 298sh, 352sh, 392
Fisetin (3,7,3′,4′)	248, 262sh, 319, 362	252, 292, 341 (dec)
Herbacetin 8-methyl ether (3,5,7,8,4′)	259sh, 276, 327, 377	289, 338, 430 (dec)
Quercetin (3,5,7,3′,4′)	255, 269sh, 301sh, 370	247sh, 321 (dec)
Quercetin 3-*O*-glucoside	257, 269sh, 299sh, 362	272, 327, 409
Quercetin 7-*O*-rhamnoside	256, 269sh, 372	291, 367, 457 (dec)
Quercetin 3,7-*O*-glucoside	256, 268sh, 355	268, 300sh, 396
Rhamnetin (3,5,7,3′,4′)	256, 270sh, 295sh, 371	242, 286, 331, 432 (dec)
Isorhamnetin (3,5,7,3′,4′)	253, 267sh, 306sh, 370	240sh, 271, 328, 435 (dec)
Patuletin (3,5,6,7,3′4′)	258, 272sh, 293sh, 371	251sh, 336
Gossypetin (3,5,7,8,3′,4′)	261, 276, 339, 385	251, 287, 366 (dec)
Myricetin (3,5,7,3′,4′,5′)	254, 272sh, 301sh, 374	285sh, 322, 423 (dec)
Isoflavones		
Daidzein (7,4′)	238sh, 249, 260sh, 303sh	259, 289sh, 328
Formononetin (7,4′)	240sh, 248, 259sh, 311	255, 273sh, 335
Genistein (5,7,4′)	261, 328sh	276, 327sh
Afrormosin (6,7,4′)	258, 320	258, 349
Tectorigenin (5,6,7,4′)	267, 330	278, 328
Orobol (5,7,3′,4′)	262, 294sh, 338sh	269, 334 (slow dec)
Flavanones, dihydroflavonols		
Pinocembrin (5,7)	289, 325sh	245, 324
Liquiritigenin (7,4′)	276, 312	250, 298sh, 327sh, 335
Naringenin (5,7,4′)	289, 326sh	245, 323
Eriodictyol (5,7,3′,4′)	289, 324sh	246, 324
Hesperetin (5,7,3′,4′)-7-*O*-rhamnoglucoside	283, 326	242, 286, 356
Dihydrokaempferol (3,5,7,4′)	291, 329sh	246, 325
Dihydroquercetin (3,5,7,3′,4′)	290, 327sh	246sh, 326 (slow dec)
Chalcones	EtOH (λ_{max}nm)	NaOEt (λ_{max}nm)
2′,4′-Dihydroxy-	267, 317, 345sh	279, 300, 394
4,4′-Dihydroxy-	240, 348	250, 427
2′,4′-Dihydroxy-4-methoxy-	237, 307sh, 362	234sh, 282, 400
Butein 4′-*O*-glucoside (2′,4′,3,4: Coreopsin)	245, 265, 305sh, 385	450

Aurones		
Sulphuretin (6,3′,4′)	257, 270, 399	464 (dec)
6-glucoside	257, 277, 340sh, 404	515
Aureusidin (4,6,3′,4′)	254, 269, 336, 399	(dec)
Maritimetin (6,7,3′,4′)	270, 355sh, 413	(dec)
6-glucoside	242, 274, 330, 419	505
Bracteatin (4,6,3′,4′,5′)	260, 327, 403	(dec)
4-glucoside	259, 330, 409	(dec)
6-glucoside	263, 322, 408	470

Anthocyanins	MeOH-HCl (λ_{max}nm)
Pelargonidin (3,5,7,4′)-	
5-glucoside	— 513
7-glucoside	270, 508
3-glucoside	270, 506
3,5-diglucoside	269, 504
3,7-diglucoside	279, 498
Cyanidin (3,5,7,3′,4′)-	
3-glucoside	274, 523
Delphinidin (3,5,7,3′,4′,5′)-	
3-glucoside	276, 534
Malvidin (3,5,7,<u>3</u>′, 4′,<u>5</u>′)-	
3-glucoside	276, 534

[a] Oxygenation pattern in parentheses; site of methylation underlined; (dec) = decomposition of sample evidenced by decreasing band intensity with time.

[b] Sources of data: Mabry *et al.* (1970), Harborne (1967), Jurd and Horowitz (1961), personal records.

4

Hydrolysis and the Analysis of Glycosides

Once individual flavonoids have been isolated by chromatography and the structural significance of spot colour, R_f and u.v.-visible absorption spectra properly assessed, further structure elucidation of the glycosides is achieved by attempting to cleave the sugar from the aglycone by hydrolysis. In this way the various glycoside types are distinguished from one another and, in cases where cleavage occurs, the liberated sugar(s), aglycone, acyl functions etc. may be separated and identified. Three types of hydrolytic treatment are commonly used for this purpose, acidic, enzymic and alkaline.

4.1 Acid treatment

The time taken for cleavage of a sugar from a flavonoid *O*-glycoside by acid treatment is determined not only by the strength of the acid, but also by the nature of the sugar (e.g. glucuronide > glucoside = galactoside > rhamnoside) and by the siting of the sugar on the flavonoid nucleus (e.g. 7-*O*-glycosides > 4′-*O*-glycosides > 3-*O*-glycosides). Thus under standard hydrolysis conditions (details below, Harborne, 1965), flavonol 3-*O*-glycosides and 7-*O*-rhamnosides hydrolyse completely in 2–6 minutes, flavone (or flavonol) 7- and 4′-*O*-glycosides and anthocyanidin 3-*O*-glycosides hydrolyse in 8–30 minutes, flavonol 3,7 and 4′-*O*-glucuronides hydrolyse in 60–250 minutes and flavonoid *C*-glycosides remain unhydrolysed.

4.1.1 *Standard procedure for hydrolysis of O-glycosides*

Dissolve the flavonoid glycoside (1 mg) in 5 ml of 2 N HCl:MeOH (1:1) in a 25-ml round bottom flask and heat (under a condenser) on a steam bath for 60 minutes. Evaporate to dryness on a rotary evaporator. Dissolve residue *completely* in a minimum of MeOH:H_2O (1:1) and chromatograph the product (p.c. or t.l.c.-cellulose, 15% HOAc) beside the starting material to determine whether hydrolysis has taken place.

a. If hydrolysis is shown to have occurred, e.g. by a reduction in R_f, the flavonoid is an *O*-glycoside (or less likely a bisulphate or *O*-glycosylated *C*-glycoside).
b. If no hydrolysis has occurred the glycoside is a *C*-glycoside or possibly a glucuronide (see section 4.1.5).
c. If partial hydrolysis has occurred the glycoside is likely to be a glucuronide (see section 4.1.5). Investigate by enzymic hydrolysis (see section 4.2).

To isolate the sugars and aglycone for further analysis, evaporate the aq. MeOH solution to half volume to remove the MeOH, then extract several times with EtOAc (by shaking vigorously in a test-tube). The aglycone will be in the EtOAc fraction and the sugars in the H_2O (but for glucuronic acid see section 4.2).

4.1.2 *Modified procedures for mild hydrolysis and acid labile aglycones*

For partial hydrolysis of a flavonoid polyglycoside the use of milder conditions, e.g. 0·2 M HCl:MeOH (1:1), 100°C, 10 min, or 1 M trifluoroacetic acid (TFA), reflux 30 min is often effective. Under these conditions, 7-*O*-diglycosides will partially be converted to 7-*O*-monoglycosides but most 3-*O*-diglycosides (unless they are glycosides of 3-*O*-glucuronides) will hydrolyse completely. Trifluoroacetic acid has the advantage that it is evaporated readily (b.p. 72·4°C) thus minimizing additional hydrolysis during the subsequent rotary evaporation. If HCl is used, however, neutralization with NH_4OH prior to evaporation will prevent further hydrolysis taking place on work-up. Alternative methods for partial hydrolyses using acids have been described by Chandler and Harper (1961), although clearly, selective enzymic hydrolysis (q.v.) must always be considered.

If the aglycone produced by the "standard procedure" is acid-labile (e.g. some flavonols) contact with acid must be minimized. Thus enzymic hydrolysis is ideal, but if acid is used the hydrolysis period should be minimized and TFA (2 M TFA: MeOH, 1:1, refluxed for 30–120 min) or HCl (subsequently neutralized) used. The use of 2 M TFA with or without MeOH may also prevent the production of Wessely–Moser rearrangement products (see section 4.1.5).

4.1.3 *Sugar analysis by paper chromatography*

Analysis by p.c. is capable of separating all commonly encountered sugars (Harborne, 1973a; Jarvis and Duncan, 1974). First a standard mixture of sugars is made up for reference purposes comprising 100 mg each of glucose, galactose, rhamnose, xylose and arabinose dissolved in 20 ml of water (containing 10% isopropyl alcohol as preservative). Aliquots (4 μl) of this reference solution are then applied at points along the baseline of a sheet of Whatman No 1 chromatography paper (57 cm length), and between these points the unknown sugars are applied as spots with a variety of different loadings (e.g. 1, 2 and 4 applications per spot). Several chromatograms should be prepared to enable development in a range of different solvents. Solvents most used include: BAW, BEW, phenol, EPAW and BBPW (see instructions, Table 2.2). The paper chromatograms are developed in the descending mode for 18–24 h (see Table 2.2) after which they are dried in a fume cupboard before being sprayed for spot detection.

Spray reagents

1. *Aniline hydrogen phthalate* Dissolve aniline (0·92 ml) and phthalic acid (1·6 g) in *n*-BuOH (49·0 ml), ether (49·0 ml) and H_2O (2·0 ml).
2. *p-Anisidine hydrochloride* Dissolve *p*-anisidine-HCl (1 g) in MeOH (10 ml), add *n*-BuOH (100 ml) and sodium dithionite (0·1 g).
3. *Diphenylamine/aniline/*H_3PO_4 (Unlike 1 and 2 above, reacts with keto-sugars, e.g. fructose, sucrose). To 1 part 85% orthophosphoric acid add 10 parts of a solution in acetone (100 ml) of diphenylamine (1 g) and aniline (1 g) just prior to spraying.

To render sugar spots visible, one of the above sprays is applied to the chromatogram which is then heated in an oven at 100–120°C for about 3 minutes or until brownish spots appear. Viewing under u.v. light (366 nm) may enhance sensitivity. In decreasing order of mobility, the sugar sequences in selected solvents are as follows:

BEW rhamnose (glucuronic acid lactone)[a], xylose, arabinose, glucose galactose, glucuronic acid

BAW rhamnose (glucuronic acid lactone)[a], xylose, arabinose, glucuronic acid, glucose/galactose

BBPW (glucuronic acid lactone)[a], rhamnose, xylose, arabinose, glucose, galactose, glucuronic acid

Phenol rhamnose (glucuronic acid lactone)[a]/arabinose, xylose, galactose, glucose, glucuronic acid

Quantification of sugar levels by this method has been described (Mabry *et al.*, 1970; Harborne, 1973a) but in the author's experience it is difficult to carry out successfully and is only applicable to mono- and di-glycosides.

[a] Acid hydrolysis treatment converts glucuronic (but not galacturonic) acid to its lactone (high R_f).

G.l.c., h.p.l.c., partial hydrolyses and occasionally ^{13}C-n.m.r. spectroscopy are alternative and often preferable procedures.

4.1.4 *Sugar analysis by gas–liquid chromatography*

Gas–Liquid Chromatography (g.l.c.) provides a more sensitive and quantitative method of sugar analysis than does p.c. However, for this method the sugars must first be converted to their trimethylsilyl (TMS) ether derivatives to render them sufficiently volatile. The minimum amount of sugar required (0·25–0·5 mg) is approximately that produced by hydrolysis of 8 ml of a glycoside solution with absorbance (A) = 1 (see Chapter 3).

Trimethylsilylation method The sugars (*c*. 0·25 mg) in a 5-ml round bottom flask are dissolved in dry pyridine (0·1 ml) and hexamethyldisilazane (0·05 ml) added by syringe. Trimethylchlorosilane (0·05 ml) is then added and the flask stoppered and left for 30 minutes. The mixture is then taken to dryness on a rotary evaporator and the residue washed with dry heptane (0·1 ml). The heptane-soluble TMS ethers should then be stored in an air-tight vial as they will hydrolyse slowly on contact with atmospheric moisture. Two microlitres of this solution should give a sufficiently intense g.l.c. trace (at attenuation 1280).

Column conditions 3% OV-1 on silanized Chromosorb W; column temperature, 180°C; injection port and detector temperature, 275–300°C; helium flow rate, *c*. 35 ml min^{-1}; flame ionization detector.

Retention times Under the above conditions and relative to *β*-glucose, retention times of acid-treated sugars are as follows (J. W. Wallace, personal communication):

Glucose	1·0 (*β*)	0·66 (*α*)
Galactose	0·69	0·57
Rhamnose	0·28	0·21
Arabinose	0·22	0·20
Xylose	0·36	0·29
Glucuronic acid	0·42 (= Glucuronolactone)	
Glucuronic acid	1·22, 0·90, 0·69 (no acid treatment)	
Galacturonic acid	0·83 (1·09, 0·59) (no acid treatment)	

Numerous peaks additional to those representing the sugars may be apparent when working with small sugar samples. These may originate from solvents or reagents used in the preparative process, or from the paper from which the original flavonoid glycoside was extracted. For this reason it is wise always to prepare a blank for analysis, consisting of the extract from blank spots cut from the original p.c., which is hydrolysed and derivatized in an identical manner to the actual sample.

4.1.5 *Flavonoid C-glycoside/O-glucuronide recognition*

If the use of the standard hydrolysis procedure (4.1.1) indicates the presence of a glucuronide or *C*-glycoside by failing to produce an aglycone, the glycoside should be checked for its t.l.c. mobility on cellulose in H_2O (be careful to exclude any traces of acid). Glucuronides are recognizable by their high mobilities (mono-, R_f *c.* 0·5; di-, R_f *c.* 0·9) relative to other glycosides (R_f 0·0 to *c.* 0·25). The glycoside should also be treated with 2M HCl at 100°C for 2 h. This will convert a glucuronide predominantly to its aglycone (degrading much of the glucuronic acid in the process), but will leave a *C*-glycoside unhydrolysed. However, a 6- or 8- mono-*C*-glycoside or an unsymmetrically substituted, 6,8-di-*C*-glycoside will undergo a Wessely–Moser rearrangement (see Fig. 4.1) resulting in the production of two major isomers which are clearly separable on cellulose t.l.c. in 15% HOAc. Symmetrically substituted di-*C*-glycosides (which have the same sugar at C-6 and C-8) will be unaffected by this treatment.

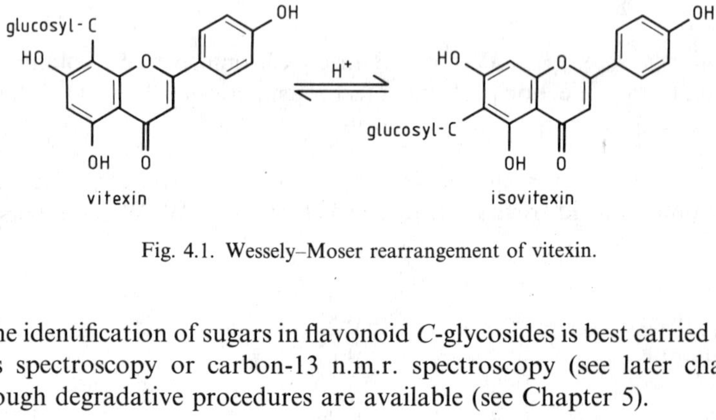

Fig. 4.1. Wessely–Moser rearrangement of vitexin.

The identification of sugars in flavonoid *C*-glycosides is best carried out by mass spectroscopy or carbon-13 n.m.r. spectroscopy (see later chapters) although degradative procedures are available (see Chapter 5).

4.1.6 *Bisulphate detection*

Flavonoid bisulphates are often recognizable from the arrow-shaped spots they form on p.c., and they are generally more readily acid hydrolysed (0·1 M TFA, 100°C, 5 min) than are *O*-glycosides. The standard hydrolysis procedure (Section 4.1.1) will thus cleave them, the sulphate ion remaining in the aqueous layer after EtOAc extraction. The presence of sulphate in this layer is indicated if, on the addition of a few drops of 2 M $BaCl_2$, a white precipitate (or cloudiness, with weak solutions) of $BaSO_4$ appears. For confirmation, t.l.c. against HSO_4^- on cellulose using 20% 0·1 M HCl in EtOH may be used (Williams *et al.*, 1973). When sprayed with sodium cobaltous hexanitrite, HSO_4^- appears as a white spot, R_f 0·72, on a pale yellow background. Additional confirmation may be obtained by paper electrophoresis (Section 2.5.2).

4.2 Enzymic hydrolysis

Enzymic hydrolysis is a useful method for establishing the nature of the linkage of a sugar to a flavonoid (i.e. *α* or *β*). Also, in theory, it offers a method for the cleaving of specific monosaccharides from flavonoid *O*-glycosides, and so identifying them. In practice, however, commercial enzyme preparations are rarely sufficiently pure to provide this specificity and it is usually advisable to carry out a sugar analysis (p.c. or g.l.c.) on the product(s) of enzyme treatment. Further, only a limited range of suitable enzymes are commercially available. Useful enzymes encountered by the author are listed below.

β-Glucosidase (emulsin)—hydrolyses *β*-D-glucosides (and xylosides) but not anthocyanidin glycosides; available from Koch–Light.
β-Galactosidase—hydrolyses *β*-D-galactosides; available from British Drug Houses.
β-Glucuronidase—hydrolyses *β*-D-glucuronides; available from Sigma and Koch–Light.
Pectinase—hydrolyses *α*-D-polygalacturonides (and *α*-L-rhamnosides); available from Koch–Light and Sigma.
Anthocyanase—hydrolyses most anthocyanidin glycosides; available Rohm & Haas Co., Pennsylvania.
Rhamnodiastase—removes many oligosaccharides intact from glycosides; from *Rhamnus frangula* seeds (Chandler and Harper, 1961).
Naringinase—hydrolyses naringenin 7-*O*-neohesperidoside to naringenin; available from Sigma.
Takadiastase—hydrolyses *α*-L-rhamnosides.

Enzymic hydrolyses normally occur rapidly, most being complete in under 1 hour. Note, however, that acylated sugars and *C*-glycosides are resistant to enzymic hydrolysis. Although it is often recommended that enzymic hydrolyses be carried out in acidic buffers of about pH 5, distilled water is generally satisfactory with flavonoid glycosides and has the advantage that it does not complicate the work-up. More detailed comments on selected enzyme preparations are presented below.

4.2.1 *β-Glucosidase (Koch-Light, from sweet almonds)*

Method To the sample (0·5 mg) in distilled water (0·5 ml) add a small amount (*c.* 6 "flakes") of enzyme. Leave for 15 minutes at 28°C (30–60 minutes at 18°C) before reducing to dryness *in vacuo*. Dissolve the residue *completely* in MeOH/H_2O and chromatograph on cellulose (p.c. or t.l.c.) in 15% HOAc.

A lowering of the mobility relative to the starting material indicates that some hydrolysis has taken place, although this should always be checked by cochromatographing starting material with product (i.e. by applying both to the same spot), as the presence of the enzyme in the product may retard migration. Under these conditions β-glucosidase will also partly hydrolyse β-D-glucuronides and β-D-galactosides and so chromatographic confirmation of the sugar(s) produced is advisable. β-Glucosidase will not, however, remove glucose from *O*-diglycosides such as rhamnoglucosides, and gluco (1–2) glucosides, but will slowly hydrolyse gluco (1–6) glucosides. If sample solubility proves to be a problem, MeOH (up to 30% by volume) may be added without markedly affecting the enzyme.

4.2.2 *Pectinase (Koch-Light, from Aspergillus niger)*

The method used is as described for β-glucosidase but the time for completion of hydrolysis varies depending upon the sugar being cleaved. Since this pectinase possesses β-glucosidase, β-glucuronidase and α-rhamnosidase activity (Markham *et al.*, 1976), 7-*O*-β-D-glucosides and glucuronides are also hydrolysed to some extent (*c.* 30%) under the conditions described above. Flavonoid 3- and 7-*O*-α-L-rhamnosides, however, require overnight treatment (at 18°C) for hydrolysis, and removal of the rhamnose from a rutinoside requires 1–3 days' treatment (but note: glucose may also be removed). Neohesperidosides appear to be unaffected by this enzyme.

4.2.3 *Anthocyanase (Rohm & Haas Company)*

This enzyme requires purification before use (Harborne, 1965). It is capable of removing a wide variety of sugars, both α- and β-linked, mono- and disaccharides from anthocyanins and flavonols. The hydrolyses are carried out at 37°C in either a lactate buffer (pH 3·95) or an acetate buffer (pH 4·0) and take from 1 to 48 hours depending upon the type of glycoside (Harborne, 1965).

4.2.4 *β-Glucuronidase (Koch-Light ex marine mollusc)*

Using the method outlined above for β-glucosidase (28°C, 30 min), both

O-glucosides and *O*-glucuronides are completely hydrolysed (for distinction, see under "acid hydrolysis"). This β-glucuronidase will also hydrolyse β-galacturonides given time (Markham, 1972). For subsequent g.l.c. identification of the glucuronic acid liberated, it is recommended that the product of enzymic hydrolysis be treated with acid (2 N HCl, 20 min, 100°C) to convert the glucuronic acid to the readily identifiable glucuronolactone (see section 4.1.4). In any case an enzyme blank should always be run.

4.3 Alkaline hydrolysis

4.3.1 *Hydrolysis of glycosides*

Alkaline hydrolysis is rarely used with flavonoid glycosides. However it does offer a method for selectively removing a sugar (or sugars) from the 7- (or 4′-) hydroxyl group in the presence of one attached to the 3-hydroxyl. This selectivity is the reverse of that with acid. Alkaline treatment will remove disaccharides from the 7-hydroxyl providing the interglycosidic linkage is not (1→2). Thus, rutinosides are hydrolysed but 7-*O*-apio(1→2)glucosides and 7-*O*-neohesperidosides are not. Care must be taken to exclude air from the reaction mixture as many flavonoids oxidize in the presence of alkali and oxygen. The method used by Litvinenko and Markov (1969) is as follows:

Heat the glycoside (10–30 mg) dissolved in 0·5% KOH (10 ml) on a steam-bath for 30 mins in a N_2 atmosphere. Neutralize with 2 M HCl and chromatograph on paper in 15% HOAc to isolate the flavonoid product.

Although most 7- and 4′-*O*-glycosides are cleaved in 30 minutes, some may require up to 2 hours. The specific removal of 4′-linked sugars in the present of 7-linked sugars has been claimed (Markham *et al.*, 1978) using modified conditions.

4.3.2 *Removal of acyl groups*

Organic acids attached to the sugar moiety of a glycoside by an ester linkage (acyl groups) may be removed by treatment with cold alkali. A convenient micro-method for carrying out this deacylation in the absence of air has been described (Markham *et al.*, 1978).

Method Mix a concentrated aqueous (or MeOH/H_2O) solution of the acylated glycoside (2 μl) with 2 M NaOH (5 μl) using a 10-μl syringe and leave the mixture in the syringe (with all air expelled) for 2 hours at room temperature. Then eject mixture into a vial containing 2 M HCl (6 μl) and evaporate the acidified solution to dryness under vacuum. The liberated acid is then extracted from the residue with warm ether. The remaining residue will contain the deacylated glycoside which may be dissolved out in MeOH and purified by p.c. or t.l.c.

To check for the occurrence of deacylation subsequent to this treatment, the relative mobilities of starting material and product should be compared by p.c. or t.l.c.—cellulose in both TBA (or BAW) and 15% HOAc. Deacylation normally reduces mobility of the flavonoid in both solvents (unless the acyl group increases the polarity of the flavonoid, e.g. caffeoyl).

For identification of the acid liberated, t.l.c. or p.c. comparison with authentic acids is recommended using the solvents: *n*-BuOH : formic acid : H_2O (4:1:5), >12 hours old, and NH_4OH:EtOH (5:95). Relative R_f values for a wide variety of organic acids have been determined for each of these solvents (see below). If the acid spot is not visible in u.v. or u.v./NH_3 it will be necessary to use a spray reagent for its detection and an excellent one for this purpose has been described (Paskova and Munk, 1960). Two solutions are made up: (a) 0·075% bromocresyl green plus 0·025% bromophenol blue in absolute EtOH and (b) 0·5% $KMnO_4$ plus 1% $Na_2CO_3.10H_2O$ in H_2O. These are mixed in a 1:1 ratio just prior to use. The t.l.c. plates must be thoroughly dried before spraying. Although the mixture is stable only for 5–10 minutes, solutions (a) and (b) may be stored for long periods. Alternatively, a universal pH indicator may be used but this would be less sensitive than the above spray.

A selection of the more useful relative (i.e. to glycolic acid) R_f values for acids reported by Carles *et al.* (1958) is as follows (NH_4OH solvent, BuOH solvent): glycolic (100, 100), lactic (145, 120), maleic (45, 105), malonic (30, 110) succinic (50, 125), tartaric (30, 35–50), benzoic (200, 160) *p*-hydroxybenzoic (140, 153), caffeic (90, 135), cinnamic (200, 160), *o*-coumaric (160, 155), gallic (70, 95), gentisic (180, 153), phthalic (60, 145), protocatechuic (90, 135), salicylic (235, 160).

If the acyl group is located on the flavonoid nucleus, its position of attachment will generally be evident from a comparison of the u.v.-visible absorption spectra before and after deacylation. However, if these spectra are identical, the acyl function is probably attached to the sugar moiety, in which case carbon-13 n.m.r. spectroscopy is the ideal method for identification of the acylated site (see section 6.2.2). Occasionally enzyme treatment of an acylated flavonoid polyglycoside will help to define the particular sugar which is acylated. This sugar will not be cleaved from the flavonoid by enzymic hydrolysis until after deacylation (see section 4.2).

4.4 Summary

As a summary to this section on hydrolysis of flavonoid glycosides by acids, enzymes and alkali, the potential for structure elucidation is exemplified by considering the effect of each on the hypothetical natural product, kaempferol 3-*O*-(2″-*O*-acetyl-rutinoside)-7-*O*-glucoside (see Fig. 4.2).

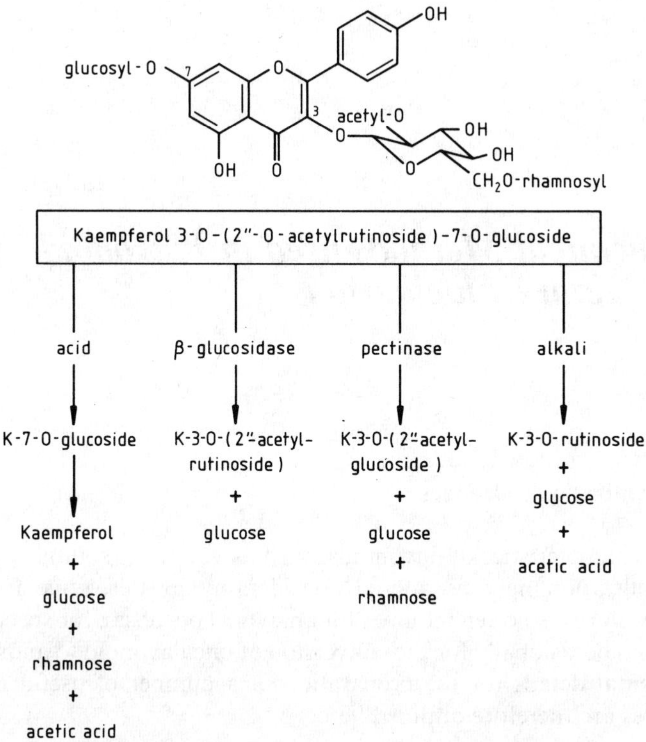

Fig. 4.2. The effects of different hydrolysis conditions on Kaempferol 3-*O*-(2″-*O*-acetyl-rutinoside)-7-*O*-glucoside.

5

Chemical Methods used in Flavonoid Structure Elucidation

5.1 Derivatization techniques

Physical characteristics of flavonoids, such as volatility, solubility and ease of crystallization may be altered considerably by derivative formation. Derivatives are also often required for analytical purposes, for spectroscopic studies and occasionally for the conversion of one flavonoid to another. Key experimental details for the preparation of a number of useful flavonoid derivatives are therefore outlined here.

5.1.1 *Trimethylsilyl ether derivatives*

Scheme 5.1 $$R-[OH]_x \longrightarrow R-[OSi(CH_3)_3]_x$$

A method for the preparation of TMS ether derivatives (Scheme 5.1) has already been described for sugars in Chapter 4, and the same procedure is also satisfactory for flavonoids and their glycosides (for a description of alternative methods, see Poole, 1977). However, since flavonoid TMS ethers are mainly prepared for ^{1}H-n.m.r. studies, the solvent used for extracting the TMS ether from the reaction flask should be spectral grade CCl_4 and not heptane.

It is important to keep these derivatives dry, as even small amounts of water will cause the TMS groups to hydrolyse off. Those on the 5- and 3-hydroxyl groups are particularly sensitive. Hydrolysis of the TMS ethers, after use, can be used to regenerate the original flavonoid. This is effected by overnight treatment with a mix of MeOH : H_2O : HOAc (150 : 30 : 1).

5.1.2 *Acetylation*

Scheme 5.2 $R{-}[OH]_x \longrightarrow R{-}[OCOCH_3]_x$

Acetylation of all but the most sterically hindered hydroxyl groups on a flavonoid aglycone may be accomplished as described below.

Method Dissolve the flavonoid (20 mg) in pyridine (0·5 ml) and add acetic anhydride (0·5 ml). Leave in a stoppered flask overnight at room temperature. Pour the reaction mixture into water (*c.* 30 ml) with stirring and leave for 1 h. Isolate the insoluble flavonoid acetate by filtration using a small Buchner funnel (see Fig. 2.5). If filtration is unsuitable (e.g. if smaller amounts of flavonoid are used) extract the aqueous mixture with ether to remove the acetate derivatives. Purify acetate by recrystallization (see section 2.6) from $MeOH/H_2O$ or by t.l.c. on silica (q.v.).

Specific acetylation of the 5-hydroxyl group in the presence of a 7-hydroxyl is possible if perchloric acid is substituted for pyridine and the reaction carried out at -10°C using a 1 mole-equivalent of acetic anhydride (Bass, 1979). The fate of other hydroxyls has not been determined.

5.1.3 *Permethylation (and perdeuteromethylation)*

Scheme 5.3 $R{-}[OH]_x \longrightarrow R{-}[OCH_3]_x$

The process of permethylation (Scheme 5.3) brings about the methylation of all hydroxyl groups, phenolic and sugar, on the flavonoid. Such derivatives are important for use in mass spectroscopy, as flavonoids are rendered more volatile by permethylation. The process recommended for the preparation of permethyl (PM) ethers is outlined below.

Reagents
1. Dimethylformamide (or dimethylsulphoxide) dried over a 4 Å molecular sieve.
2. Powdered sodium hydride (NaH)-oil coated.
3. Methyl iodide (b.p. 42°C., keep refrigerated; caution, carcinogen).
4. A source of oxygen-free N_2.

Equipment
1. Two syringes (1-ml capacity) with detachable needles, plus one extra needle.
2. A septum-sealed vial of 3–10 ml capacity (e.g. screw-cap septum vials or Reactivials, available Pierce Chemical Company).
3. A tube (or tower) containing fresh $CaCl_2$ to dry the N_2.

Method Add excess NaH powder (i.e. *c.* 0·5 "ml") to the open vial and remove adhering oil by washing twice with dry petroleum ether, b.p. 60–80°C (use a Pasteur pipette). Fit the septum and three needles quickly (see Fig. 5.1), and admit a stream of dry N_2 via needle A for about 10 minutes to flush out O_2 and to evaporate remaining petroleum ether. Admit the flavonoid (1–2 mg), in dry dimethyl formamide (*c.* 1 ml), to the vial via syringe at B and swirl the vial to mix the reactants. Stop the N_2 flow by removing needle A and add CH_3I (0·2 ml) via syringe at B. Mix again by swirling vial, remove needle B and store vial in the dark for 1 h. Completion of the reaction is indicated by the loss of the yellow colour in the reaction mixture. Remove the septum and add dropwise with *care*! sufficient MeOH to destroy all remaining NaH. Make up volume to *c*. 5 ml with water. Extract with $CHCl_3$ (2 × 5 ml) in a small separating funnel to isolate methylated products, and wash $CHCl_3$ solution with H_2O (using the same funnel). Add anhydrous Na_2SO_4 to the washed $CHCl_3$ solution, decant the dried $CHCl_3$ and evaporate to yield a gum of methylated products.

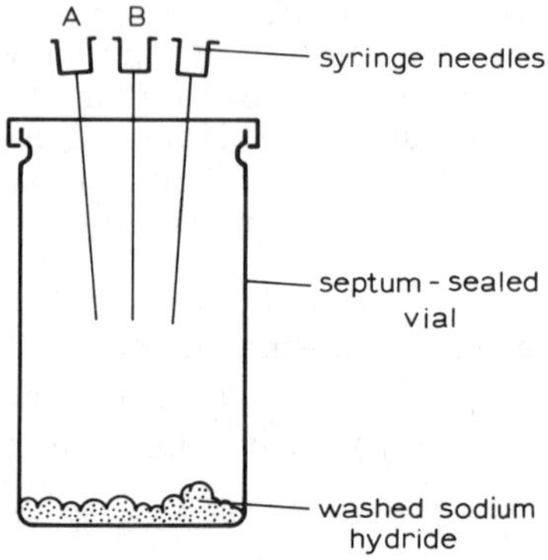

Fig. 5.1. Apparatus for permethylation reaction.

The above method usually produces a mixture of products and it is therefore advisable to isolate the PM ether pure by 1D-t.l.c. on SiO_2 using the solvent $CHCl_3$:acetone (4:1). The PM ether is normally evident at high R_f as a major bright blue fluorescent band in u.v. light (366 nm). The product or products of interest may be isolated by scraping the band(s) from the plate into a Pasteur pipette plugged at the neck with glass wool, and then eluting the PM ether from this "mini" column with a small amount of redistilled acetone. Further purification on a "mini" column of LH-20 (using MeOH) will remove SiO_2 contamination, if this is a problem.

If CD_3I is substituted for CH_3I in the above procedure, perdeuteromethyl ether derivatives are formed. These derivatives are especially useful for m.s. studies if the flavonoid contains one or more methyl groups.

5.1.4 *Partial methylation (ethylation) with diazomethane (diazoethane)*

Diazomethane (CH_2N_2) will methylate flavonoid hydroxyls at varying rates according to their acidities. The most acidic is generally the 7-hydroxyl and this can be methylated preferentially with CH_2N_2. Products with additional methylation at the other centres (3-OH and 4′-OH) are produced in only low yield if contact between the flavonoid and CH_2N_2 is kept to a minimum. The 5-hydroxyl is very difficult to methylate.

Preparation of diazomethane and diazoethane

1. According to Vogel (1961) For the preparation of diazomethane, the reaction of alkali with *N*-methyl *N*-nitroso-*p*-toluene sulphonamide (available from Merck) is utilized. All manipulations should be carried out in a fume cupboard, and plastic gloves worn, since diazomethane is known to be toxic and carcinogenic.

Method First set up a small-scale distillation unit consisting of a 100 ml distillation flask, a water-cooled condenser, and a long take-off adapter capable of delivering the CH_2N_2 to the bottom of an ice-cooled receiving test-tube containing diethyl ether (5–10 ml). Use cork joints rather than ground-glass. Add *N*-methyl *N*-nitroso-*p*-toluene sulphonamide (2·1 g) in ether (30 ml) to the distillation flask, cool in ice and then add KOH (0·4 g) in 96% EtOH (12 ml) and attach flask to condenser. After 5 minutes, distil the ethereal CH_2N_2 using a water bath. The distillate will contain about 0·3 g CH_2N_2 in ether and can be stored in a refrigerator for several weeks, if sealed in a corked (not stoppered) container with KOH pellets.

2. Using the Aldrich MNNG diazomethane apparatus This simple glass apparatus (available from Aldrich Chemical Co., Milwaukee, Wisconsin, at *c.* $30, 1980) enables the preparation of ethereal diazomethane in small quantities without the need to distil. The action of 50% aq. KOH on MNNG (*N*-methyl-*N*′-nitro-*N*-nitrosoguanidine) is used to generate diazomethane and the *N*-ethyl equivalent is used if diazoethane (for ethylation) is required.

Use of diazomethane (or diazoethane)

Method Add diazomethane in ether (*c.* 1 ml) dropwise, with swirling, to an ice-cooled solution of the flavonoid (3–5 mg) in MeOH (2 ml). Leave the mixture stoppered, at 0°C for 5 minutes and then evaporate to dryness under a vacuum. A t.l.c. comparison of the product with the starting material will reveal the amount of methyl ether formed (methyl ethers are more mobile than starting material on cellulose in CAW or BAW). If the yield is low, repeat the treatment with additional CH_2N_2. 7,4′- and 3,7-Dimethyl ethers may also be formed (e.g. Markham and Williams, 1980).

5.1.5 *Methylation (or ethylation) with dimethyl (or diethyl) sulphate*

Flavones and Flavonols may be selectively methylated by use of dimethyl sulphate/acetone/$NaHCO_3$ or dimethyl sulphate/acetone/Na_2CO_3 (Simpson and Beton, 1954). With the former, the relative rates of methylation are 7-OH > 4′-OH > 3′-OH > 3-OH, while with the latter the reverse order is claimed. The 5-OH is difficult to methylate. The general method is applicable to most flavonoid types (but not anthocyanins), a typical example being as follows (Ingham, 1979).

Method 5,4′-Dihydroxy-7,2′-dimethoxyisoflavanone (10 mg) in dry acetone (5 ml), anhydrous K_2CO_3 (1 g) and dry diethyl sulphate (15 μl, dried over anhydrous K_2CO_3), were refluxed for 2 h, the top of the condenser being fitted with a $CaCl_2$-filled drying tube. Filtration, evaporation of the filtrate and chromatography (SiO_2, $CHCl_3$) gave 5-hydroxy-7, 2′-dimethoxy-4′-ethoxyisoflavanone (9 mg).

With polyhydroxylated flavonoids the reaction may be monitored by t.l.c. in order to gauge the reaction time required to achieve the desired degree of derivatization.

5.1.6 *Isopropylidene derivatives*

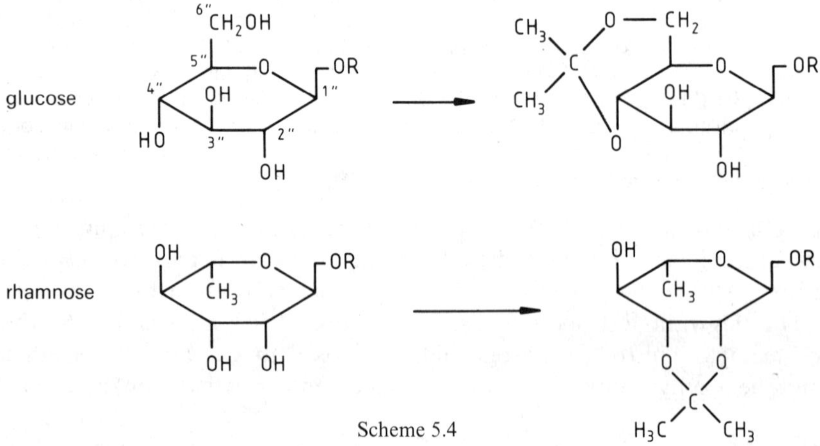

Scheme 5.4

Isopropylidene bridging of suitably positioned hydroxyl groups on the sugars of flavonoid *O*- and *C*-glycosides (e.g. Scheme 5.4) can provide useful derivatives for confirming the nature of the sugar, and defining the ring size and position of inter-glycosidic linkages by mass spectroscopy (after subsequent permethylation: Chopin, 1977; Seligmann and Wagner, 1978; Wagner and Chari, 1979). The isopropylidene derivatives are stable to alkali but readily hydrolysed by acid.

Method (i) The glycoside (1 mg) in dry acetone (10–15 ml) is stirred for 6 h with dry (anhydrous) $CuSO_4$ in a stoppered flask. The derivative is obtained from the acetone solubles by filtration and evaporation.

This method is useful for distinguishing galactosides (which derivatize completely in 6 h and yield two different derivatives; 4″, 6″, and 3″, 4″ bridged) from glucosides (which take 36 h, and form only one derivative).

Method (ii) As for (i) but substitute dry *p*-toluene sulphonic acid (freshly melted) for $CuSO_4$ and stir for 15 h. Neutralize reaction mixture (to pH 7) with 10% NH_4OH before reducing to dryness.

The isopropylidene derivatives formed may be purified by t.l.c. on SiO_2 using EtOAc:MeOH:H_2O (100:17:13), or, if they are permethylated, $CHCl_3$:acetone (9:2).

5.1.7 *References to other useful derivatization techniques*

6- and 8-C-Prenylation and O-prenylation, e.g. flavonoid + prenyl bromide + NaOMe (Jain *et al.*, 1978; Jain and Gupta, 1975).
3- and 7-O-Glycosylation, e.g. derivatized flavonoid + acetobromoglucose + AgO + pyridine (Horhammer *et al.*, 1966; Farkas *et al.*, 1976; Vermes *et al.*, 1976).
6- and 8-C-Glycosylation, see summary of methods and applications (Chopin and Bouillant, 1975).
Methylenation, e.g. *o*-dihydroxy-flavonoid + CH_2I_2 + K_2CO_3 in acetone (Grover *et al.*, 1964).
6- and 8-C-Methylation, e.g. 5,7-dihydroxy-flavonoid + $NaNH_2$ + liquid NH_3 + CH_3I (Beirne *et al.*, 1975, and references therein).

5.2 Degradative techniques

5.2.1 *Demethylation (and de-C-glycosylation) methods*

Method (i) The sample (2 mg) in a small round-bottom flask is covered with phenol crystals and HI (1 ml) is added. The mixture is then refluxed (heating mantle or oil-bath) under a H_2O-cooled condenser for 6 h. H_2O (5 ml) is then added and the product extracted out with EtOAc. I_2 contaminates the product. Purify by p.c.

Method (ii) As above but the sample is refluxed for 2 h with 50:50 Ac_2O/HI (2 ml) and one crystal of phenol. $NaHSO_3$ solution is then added to neutralize the acid and the product either removed by filtration or extracted out with EtOAc. Purify by p.c.

Method (iii) The sample (1 mg) in a small test-tube is covered with solid pyridinium chloride (*c.* 0·5 ml) and a cotton plug inserted into the top of the test-tube. The mix is melted in an oil bath at 140–150°C and maintained at that temperature for 2–6 h. After cooling, H_2O is added and that product extracted out with EtOAc. Purify by p.c.

All of the above methods give reasonable yields of demethylated product but poor yields of de-*C*-glycosylated product. Selective demethylation at the 5-hydroxyl has been described by Dean (1966) using BCl_3 (or BF_3, BBr_3) in methylene chloride.

5.2.2 *Cleavage of disaccharides intact from flavonoid 3-O-diglycosides*

Method To the flavonol (or anthocyanidin) 3-*O*-diglycoside (1 mg) in a vial is added 3 drops of 2 M NH_4OH followed by 1–2 drops of H_2O_2 (100 volume). The colour disappears as the reaction proceeds, and this can take up to 2 h. Evaporate to dryness (with heat to destroy the H_2O_2), dissolve residue in H_2O and chromatograph against disaccharide standards (p.c. or t.l.c. on cellulose).

5.2.3 *Cleavage of sugar moiety intact from a flavonoid C-glycoside*

C-linked sugars are best identified by ^{13}C-n.m.r. spectroscopy (q.v.). Cleavage methods without exception give poor yields of the sugar and consume sizeable quantities of flavonoid. The aglycone portion may be isolated by use of one of the demethylation techniques (section 5.2.1).

Method (Chawla *et al.*, 1974). The *C*-glycoside (20–50 mg) and $FeCl_3$ (250 mg) in H_2O (1·5 ml) are refluxed for 6 h. The reaction mixture is then diluted to 5 ml with H_2O and the pH adjusted to 8 with NaOH. It is then filtered to remove brown insolubles and passed through a small column of SiO_2 using H_2O as eluant. The sugar(s) eluted are then identified by p.c.
Note This method yields glucose *and* arabinose from *C*-glucosides but rhamnose only from *C*-rhamnosides.

5.2.4 *Alkaline degradation of isoflavones to deoxybenzoins*

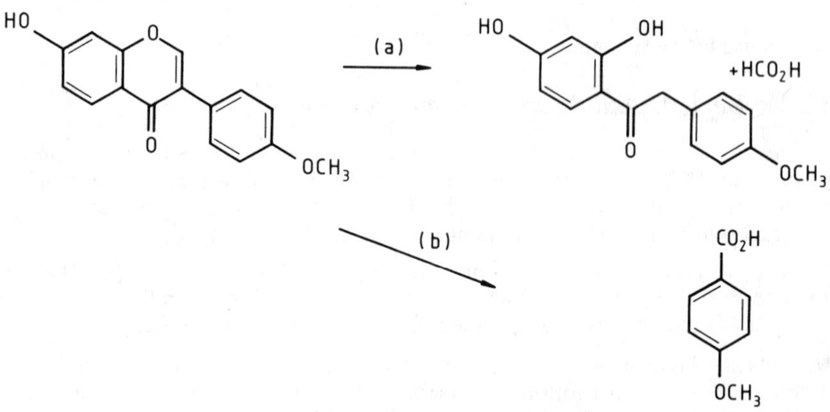

Scheme 5.5 (a, b)

This method (Scheme 5.5(a)) which is especially successful with 5-deoxyisoflavones, is used (i) to establish that the flavonoid is an isoflavone (deoxybenzoins have highly distinctive u.v.-visible absorption spectra, e.g. λ_{max} MeOH, 270, 320 nm; λ_{max} NaOMe 360 nm), and (ii) to define the oxygenation pattern via the ready mass spectrometric breakdown of the deoxybenzoin.

Method The isoflavone (2 mg) in 5% NaOH is stirred at 40°C for 1–2 h. The solution is then neutralized with HCl and the product extracted out with ether and purified by t.l.c.

5.2.5 *Alkaline oxidation of isoflavones to benzoic acids (derived from B-ring)*

This method offers a convenient means of determining the B-ring oxygenation pattern of unknown isoflavones via the production of a readily identifiable (t.l.c., m.s.) benzoic acid from the B-ring (Scheme 5.5(b) and Ingham *et al.*, 1977).

Method The isoflavone (or isoflavanone) must first be methylated or ethylated to ensure that all B-ring hydroxyls are derivatized. The derivatized isoflavone (2 mg) is then dissolved in EtOH (2 ml) and 15% aq. KOH (0·5 ml). Then 30% H_2O_2 (0·6 ml) is added dropwise with stirring over a period of 1 h. Stirring is continued for a further 18 h (at room temperature) and the solution subsequently acidified with 2 M HCl and extracted with EtOAc. The benzoic acid in the EtOAc is then identified by t.l.c. comparison with suitable standards (e.g. SiO_2; *n*-pentane:ether;HOAc, 75:25:3 or $CHCl_3$:MeOH, 50:1).

5.2.6 *Alkaline cleavage of flavonoids (general)*

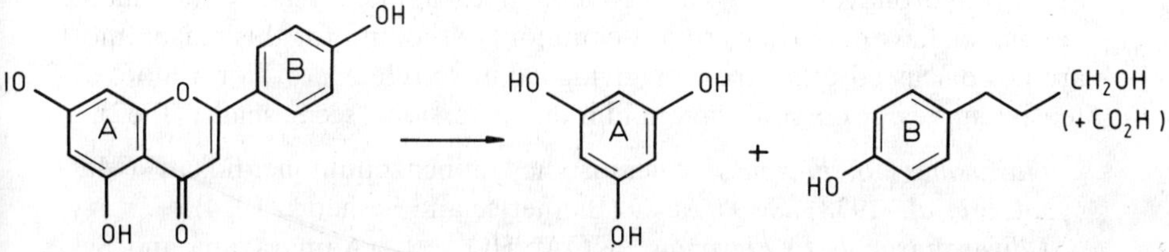

Scheme 5.6

The technique described here was developed by Hurst and Harborne (1967) to improve upon the classical alkali fusion method of cleaving the central pyran ring and producing identifiable A- and B-ring fragments. It is milder than the fusion method, is carried out in a reducing atmosphere, works well on a microscale and is applicable to a wide range of flavonoid types (see flavone example, Scheme 5.6).

Method 2 M NaOH (20 ml) in a 250-ml 3-necked flask fitted with a N_2 inlet tube and condenser is heated to 95°C and degassed by passing O_2-free N_2 through it for *c.* 15 min. 2% sodium/mercury amalgam (15 g) is added and the O_2-free N_2 atmosphere maintained. The sample is then added (*c.* 5 mg, which may be in the form of p.c. spots *on* paper!) and the heating and N_2 flushing continued for $1\frac{1}{4}$ h. The colourless reaction mixture is then cooled, concentrated HCl (5 ml) added, and ether extracted to remove products. Recommended analysis is by 2 D-t.l.c. on SiO_2 using (i) 10% HOAc in $CHCl_3$ and (ii) 45% EtOAc in Bz. Relative mobilities of commonly encountered products are detailed by Hurst and Harborne (1967). Phenolic compounds are detected as blue spots on a yellow background by spraying with Folin-Ciocalteau reagent and fuming with NH_3.

It is worth noting that the classical alkaline method of cleaving the central pyran ring is of particular value with biflavonoids. These compounds produce three types of fragment, two derived from aromatic rings not involved in the interflavonoid link and one derived from the two aromatic rings involved in the interflavonoid link. Important information regarding the interflavonoid linkage points may be gained from this data (e.g. Jackson *et al.*, 1971; Karanjaokar *et al.*, 1967).

Much of the information gained in early years from alkaline cleavage reactions is now more conveniently obtained from studies of the mass spectra or n.m.r. spectra of the unknown compounds (q.v.). For this reason, the commital of significant amounts of a flavonoid to alkaline cleavage should only be done after due consideration of all available m.s. and n.m.r. data.

5.2.7 *Potentially useful interconversions of flavonoids types*

The methods below are generally employed to facilitate the identification of a 2,3-dihydroflavonoid by its conversion to a more readily identifiable flavone or flavonol. They are not commonly used and for this reason most are not discussed other than by giving example references. For a more extensive survey of possible flavonoid interconversions, see Seshadri (1962).

Flavanones to Flavones Dichlorodicyanobenzoquinone/dioxan (Matsuura *et al.*, 1978); SeO_2/Ac_2O (Bannerjee and Seshadri, 1954).
Dihydroflavonols to Flavonols $KOAc/HOAc/I_2$ (Adinarayana and Seshadri, 1965).
Isoflavanones to Isoflavones SeO_2/Ac_2O or SeO_2/amyl alcohol (Narashimhachari and Seshadri, 1952).
Proanthocyanidins (Leucoanthocyanidins) to Anthocyanidins Boil proanthocyanidin in *n*-butanol containing 5% of concentrated HCl for 1–2 hours.

5.2.8 *Colour tests for flavanones and dihydroflavonols*

Conversion of flavanones and dihydroflavonols to red-coloured products by various reducing agents can provide a useful method for confirming the flavonoid type of unknowns which are suspected as being flavanones or dihydroflavonols from their chromatographic and u.v.-visible absorption data.

Method (i) Mg/HCl (Shinoda Test) Dissolve a few crystals of the flavonoid in a drop or two of EtOH, add Mg powder and then a drop of 5 M HCl. View against a white background. Both flavanones and dihydroflavonols turn a deep-red or magenta immediately.

Method (ii) Zn/HCl (Pew, 1948) If Mg is replaced by Zn in the above test, only dihydroflavonols give the deep-red to magenta colour. Flavanones (and flavanols) give either no colour or weak pink to lavender colours.

This method has been adapted for paper and thin-layer chromatography by Barton (1968). Zinc dust is brushed on to the surface of the p.c. or t.l.c. (cellulose) or sprayed on as a suspension in acetone, and the chromatogram then sprayed with 6 M HCl to detect dihydroflavonols. With thin-layer chromatograms, the Zn dust may be incorporated into the cellulose layer (2% by weight) providing acid solvents are not used in the subsequent chromatographic work.

Method (iii) NaOAc/Ac$_2$O/HCl (Pacheco's Test) Heat the unknown (1 mg) over a flame, first with a mixture of a few crystals of NaOAc and Ac$_2$O (0·1 ml) and then with concentrated HCl (0·1 ml). A red colour is developed ("flavanonol red", Kubota *et al.*, 1966) only with dihydroflavonols. Flavones, chalcones, aurones, flavonols and flavanones all gave a negative response, but catechins and proanthocyanidins were not tested.

6

The Usefulness of n.m.r. and m.s. Data in Flavonoid Structure Elucidation

In this chapter, the usefulness of ^{1}H- and ^{13}C-nuclear magnetic resonance (n.m.r.) spectroscopy and mass spectroscopy (m.s.) in flavonoid structure determination is summarized. Theoretical aspects of these techniques are not discussed, the emphasis being on discussion of the types of problem best solved by each technique, sample preparation, and interpretation of spectra. Recent reviews of each of these techniques with respect to their application to flavonoids are readily available and provide more detailed information if required. (Markham and Mabry, 1975—^{1}H-n.m.r.; Mabry and Markham, 1975—m.s.; Markham and Chari, 1982—^{13}C-n.m.r.).

6.1. Proton magnetic resonance (^{1}H-n.m.r.) spectroscopy

6.1.1 *Typical applications*

i. Definition of the oxygenation pattern (all three rings).
ii. Determination of the number (and position) of methoxyl groups.
iii. Distinction of isoflavones, flavanones and dihydroflavonols.
iv. Determination of the number of sugars present (and whether α- or β-linked).
v. Detection of hydrocarbon side chains such as C-linked CH_3 and C-(or O-) linked prenyl.

6.1.2 *Introduction to ^{1}H-n.m.r. spectroscopy*

The ^{1}H-n.m.r. spectrum appears predominantly in the range 0–10 p.p.m. downfield from the reference signal of tetramethylsilane (arbitrarily defined as occurring at 0 p.p.m.) Only protons produce signals (resonate) in this range and chemically identical protons are represented by the same signal. The signal size (integral) is proportional to the number of protons it represents. Thus, a signal representing an -OCH_3 group will integrate at three times the value of a signal representing a single aromatic proton. The signal position on this 0–10 p.p.m. scale (i.e. the "chemical shift") defines to some extent the type of proton it represents and the listing in Table 6.1 for TMS-ether derivatives of flavonoids in CCl_4 demonstrates this.

Table 6.1
Approximate chemical shifts of various flavonoid proton types[a]

Chemical shift (p.p.m.)	Proton type
0	Tetramethylsilane (reference)
0–0·5	Trimethylsilyl ether groups
c. 1·0	Rhamnose C-$C\underline{H}_3$ (broad doublet)
c. 1·7	Prenyl (-CH_2-CH=C(CH_3)$_2$) methyl groups (other protons, 3·5 and 5·2 p.p.m.)
c. 2·0	Acetate (-OCOC$\underline{H}_3$) and aromatic C-$\underline{C}H_3$
2–3	H-3 of flavanones (two proton-multiplet)
3·5–4·0	Most sugar C-H
4·2–6·0	H-1 of sugars (also H-2 dihydroflavonols, 5·0 p.p.m. and H-2 of flavanones 5–5·5 p.p.m.
c. 6·0	Methylenedioxy (O-C$\underline{H}_2$-O), singlet
6·0–8·0	A- and B-ring protons (see later examples)
7·5–8·0	H-2 of isoflavones (singlet)
12–14	5-OH (observed only when solvent = DMSO-d_6)

[a] Figures quoted are for flavonoid TMS-ether derivatives.

^{1}H-^{1}H coupling A single proton is often represented in a spectrum by a group of signals which together integrate for one proton. This occurs when the proton has a proton neighbour or neighbours on nearby carbon atoms. The phenomenon is referred to as "coupling" or "splitting" and the extent and size of the coupling is indicative of the number and relative siting of nearby protons on the flavonoid. For example, on an aromatic ring, two protons *ortho*-related to one another would each appear as doublets (coupling constant, $J_{,}=c.$ 9 Hz) but when *meta*-related, the coupling constant is only *c.* 2.5 Hz (see Spectrum 1, apigenin). These effects are additive. Thus if a proton has both *ortho*- and *meta*-related proton neighbours (e.g. H-6 in robinetin, Spectrum 4) it will appear as a quartet of signals, $J=9$ and

2·5 Hz. Coupling is not observed when neighbouring protons are in equivalent environments, that is, with the same chemical shift (e.g. H-2′ and H-6′ in the spectrum of robinetin).

6.1.3 *Sample size and solvents*

The amount of sample required depends very much on the spectrometer to be used. Whereas with an older type (continuous wave) spectrometer a sample of 5–25 mg would be needed to produce an acceptable spectrum, the more modern Fourier Transform-equipped spectrometers are capable of producing excellent spectra with as little as 0·1 mg of sample. $CDCl_3$ is an ideal solvent for less polar flavonoid aglycones (Mabry *et al.*, 1970a), but it is not suitable for the majority of flavonoids.

DMSO-d_6 solvent Hexadeuterodimethyl sulphoxide (DMSO-d_6) is a good solvent for flavonoid aglycones and glycosides, and has the advantage, if dry, of permitting the detection of -OH proton signals. (These disappear on addition of D_2O.) The main disadvantage of DMSO-d_6 as a solvent for ^{1}H-n.m.r. studies is that its own signals (due to partly deuterated DMSO contaminants), and the signals of the H_2O it so readily absorbs, may obscure flavonoid signals in the 2–4 p.p.m. region. A compilation of flavonoid data determined with this solvent has been published by Batterham and Highet (1964) and Hillis and Horn (1965).

CCl_4 solvent—TMS ethers Many workers prefer to convert the flavonoid (or glycoside) into its trimethylsilyl (TMS) ether derivative (for preparation, see Chapter 5). TMS ethers are readily soluble in CCl_4 which, being cheap, dry and containing no protons to complicate the flavonoid spectrum, is an ideal solvent. One big advantage of obtaining ^{1}H-n.m.r. data on these derivatives is that the interpretation is aided considerably by the ready availability of large numbers of spectra of flavonoid TMS-ethers (Mabry *et al.*, 1970a). The derivatives are prepared in almost 100% yield and the original flavonoid is readily regenerated (see Chapter 5).

Anthocyanins In the special case of anthocyanins and anthocyanidins, DMSO-d_6 containing 5–10% of DCl or deuterotrifluoroacetic acid have been proved suitable solvents (Nilsson, 1973a). TMS-ether derivatives are less useful (e.g. Bombardelli *et al.*, 1976).

Solvent shifts of methoxyl signals The measurement of ^{1}H-n.m.r. spectra of a methoxylated flavonoid in two different solvents has been used to determine the positioning of methoxyl groups on the flavonoid nucleus from the shifts observed in the methoxyl signals (for review see Markham and Mabry, 1975). With highly methoxylated flavonoids the two solvents commonly used are C_6D_6 and $CDCl_3$ (e.g. Pelter and Amenechi, 1969) and with TMS-ethers,

CCl_4 and C_6D_6 (Rodriguez *et al.*, 1972). The use of lanthanide shift reagents is proving a useful alternative method for the positioning of methoxyl groups (Okigawa *et al.*, 1972; Joseph-Nathan *et al.*, 1981).

6.1.4 *Example spectra and their interpretation*

Spectra 1–6 of flavonoid TMS-ether derivatives (in CCl_4) have been selected to illustrate commonly encountered features of flavonoid ^{1}H-n.m.r. spectra. They are derived (with some modification) from the compilation of Mabry *et al.* (1970a). In each case the integrated spectrum is plotted horizontally across the resonance spectrum.

The ^{1}H-n.m.r. spectrum of a *new* flavonoid may be roughly calculated from that of a closely related known flavonoid by application of the substituent effect data reported by Ballantine and Pillinger (1967). These workers tabulated signal shifts observed for protons *ortho*-, *meta*- and *para*-related to newly introduced substituents such as —OH, $—OCH_3$, $—OCOCH_3$, $—CH_3$, $—CO_2H$, —CH=CH—R etc. Such data can be used to estimate the effect on the spectrum of the addition (or removal) of a particular substituent.

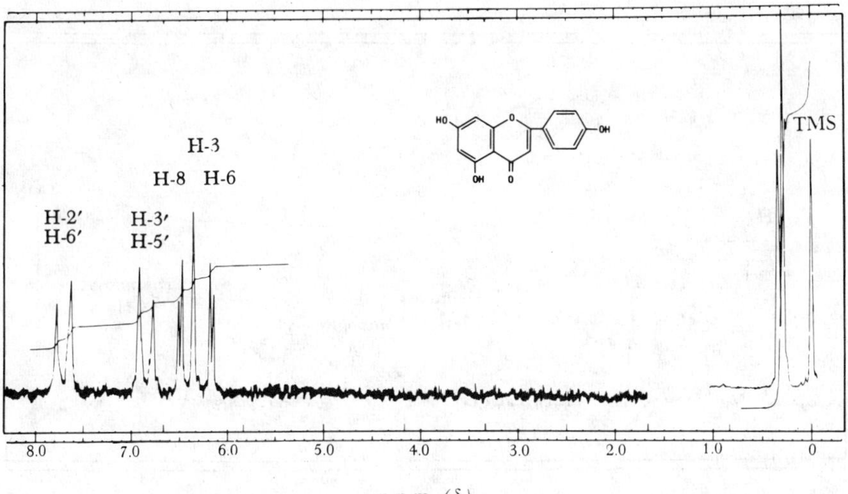

Spectrum 1. ^{1}H-n.m.r. spectrum of apigenin (TMS-ether) in CCl_4

1. *Meta*-related H-6 and H-8 appear as doublets (J=2·5 Hz).
2. H-3 with no proton neighbours appears as a singlet.
3. *Ortho*-related H-2′ and H-3′ (and H-5′, H-6′) appear as doublets (J=9 Hz).
4. H-2′ and H-6′ are identical (i.e. signals superimposed) as also are H-3′ and H-5′.

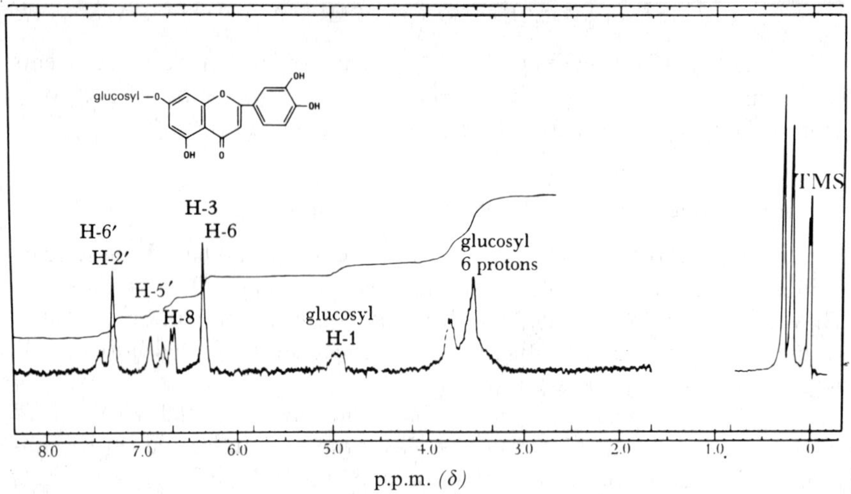

Spectrum 2. [1]H-n.m.r. spectrum of luteolin 7-*O*-glucoside (TMS-ether) in CCl_4

1. Coupling of A- and C-ring protons as for apigenin, but the effect of 7-*O*-glycosylation has caused the H-6 and H-8 doublets to move downfield by 0·2 p.p.m. and H-6 is now superimposed on H-3.
2. H-5′ appears as a doublet (J=9 Hz) due to *ortho*-coupling with H-6′.
3. H-2′ (*meta*-coupled) and H-6′ (*meta*+*ortho*-coupled) are superimposed.
4. Glucose protons are evident at 3·3–4·0 p.p.m. H-1 of glucose resonates at 5·0 p.p.m. (≡4′-*O*-glucoside but not 3-*O*-glucoside, cf. rutin, below). Combined signals integrate for 6H.

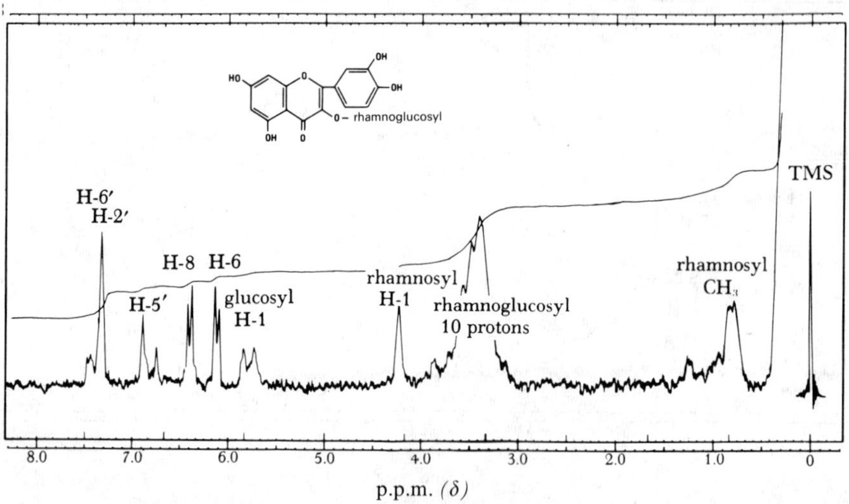

Spectrum 3. [1]H-n.m.r. spectrum of rutin (TMS ether) in CCl_4

1. Aromatic protons show pattern similar to luteolin 7-*O*-glucoside but the H-3 signal is missing.
2. Sugar protons (integrating for two hexoses) visible.
3. Rhamnose presence is indicated by C—CH_3 at 0·8 p.p.m.
4. H-1 of glucose for a 3-*O*-glucoside resonates at 5·8 p.p.m. (cf. 7-linked glucose, 5·0 p.p.m.).
5. Glucose is *β-linked* (to 3-OH). Evident from large coupling constant (6 Hz) of H-1.
6. L-Rhamnose is α-linked to glucose (H-1 a broad singlet, J=2Hz).

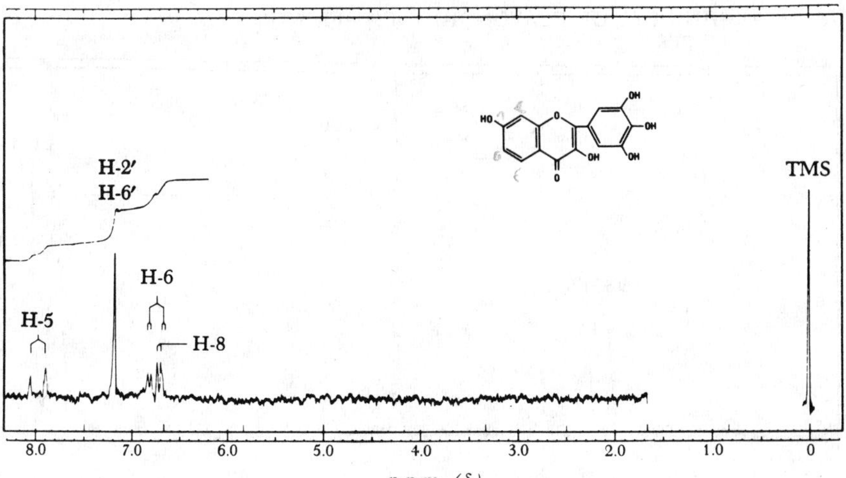

Spectrum 4. ^{1}H-n.m.r. spectrum of robinetin (TMS-ether) in CCl_4

1. H-2′ and H-6′ form one two-proton singlet (being identical they do not exhibit *meta*-coupling).
2. H-5 appears at a characteristic low field (8 p.p.m.) and is *ortho*-coupled to H-6.
3. H-6 is further coupled to the *meta*-related H-8.

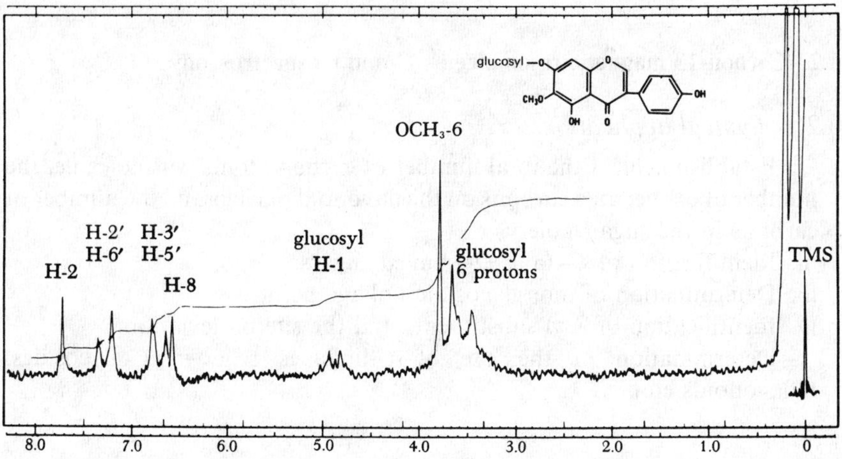

Spectrum 5. ^{1}H-n.m.r. spectrum of tectoridin (TMS-ether) in CCl_4

1. Isoflavone H-2 singlet characteristically appears at low field (cf. H-3 of flavones).
2. Methoxyl protons are clearly visible as a three proton singlet at 3·75 p.p.m.
3. H-8 appears as a singlet because C-6 is substituted (cf. apigenin).

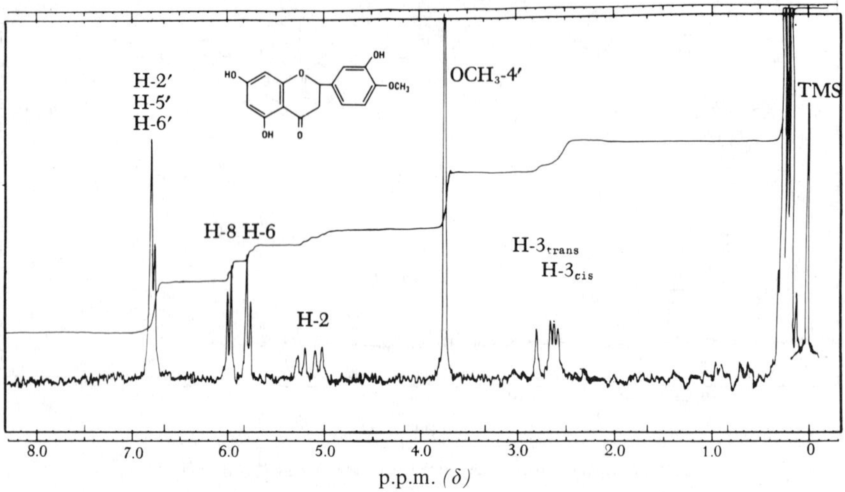

Spectrum 6. ^{1}H-n.m.r. spectrum of hesperetin (TMS-ether) in CCl_4

1. A- and B-ring patterns similar to luteolin.
2. C-ring H-2 appears as a double doublet (caused by coupling with *both* protons at C-3.
3. C-3 proton signals at 2·6–2·8 p.p.m. integrate for two protons. Superimposition and different coupling constants produce a multiplet. (In *dihydroflavonols* H-3 appears as a doublet near 4·2 p.p.m., J=11 Hz, and H-2 as a doublet near 5·2 p.p.m.)

6.2. Carbon-13 magnetic resonance (^{13}C-n.m.r.) spectroscopy

6.2.1 *Typical applications*

i. Establishment of the total number of carbon atoms per molecule, the number of oxygenated carbons on the flavonoid nucleus and the number of carbons in the sugar moiety.
ii. Identification of *C*- (and *O*-) linked sugars.
iii. Determination of interglycosidic linkage points.
iv. Identification of acyl substituents and the site of acylation.
v. Determination of the site of *C*-linkages (e.g. in *C*-glycosides, biflavonoids etc).

6.2.2 *Introduction to ^{13}C-n.m.r. spectroscopy*

The natural abundance of ^{13}C is only 1·1% and it is this 1·1% of any flavonoid sample that gives rise to the ^{13}C-n.m.r. spectrum. Carbon-13 resonance occurs predominantly in the range 0–200 p.p.m. downfield from tetramethylsilane (TMS), each different carbon being represented by one signal.

In contrast to proton resonance signals, the intensity of a carbon-13 resonance signal does not necessarily reflect the number of carbons it represents and so integration of ^{13}C-n.m.r. spectra is rarely of value.

As in ^{1}H-n.m.r. spectra, coupling between the signals of near-neighbour atoms occurs in ^{13}C-n.m.r. spectra. This is due to the interaction of the carbon with nearby protons (^{13}C-^{1}H coupling)[1] and can be most useful for signal assignment purposes. However, because ^{13}C-^{1}H coupling complicates initial interpretation of the spectrum, it is usually suppressed electronically to produce a "proton decoupled" spectrum in which every different carbon is represented by one unsplit signal. ^{13}C-^{1}H coupling constants for aromatic ring carbons range from 155–170 Hz for a directly attached proton down to 1–3 Hz for *ortho*- and *para*-related protons. A *meta*-related proton causes a 6–8 Hz splitting of the carbon signal.

Chemical shifts and substituent effects The position of a signal relative to the TMS reference (0 p.p.m.) is a good guide as to the type of carbon represented (see Table 6.2 and Figure 6.1). As indicated in Table 6.2 this position (chemical shift) is affected markedly by nearby substituents. Such effects are predictable in their extent and have led to the formulation of "substituent effect" data which defines the expected effect (on aromatic carbon resonances) of introducing a new substituent into an aromatic ring (see, e.g. Levy and Nelson, 1972; Stothers, 1972). Relevant (to flavonoids) shift data

Table 6.2
Carbon-13 chemical shift ranges for various flavonoid carbon types

Carbon type	Usual chemical shift range (p.p.m. from TMS)
Carbonyl (4-keto, acyl)	210–170
Aromatic and olefinic:	
a. oxygenated	165–155 (no o/p oxygenation); 150–130 (with o/p oxygenation)
b. non-oxygenated	135–125 (no o/p oxygenation); 125–90 (with o/p oxygenation)
Aliphatic:	
a. oxygenated (sugars)	83–69 (C-1 of *O*-glycoside, *c.* 100 p.p.m.)
b. non-oxygenated (C-2, 3 flavanones	80–40 (epicatechin C-4, 28 p.p.m.)
Methylenedioxy	*c.* 100
O-CH_3	55–63 (60–63 = *o*-disubstituted)
C-$\underline{C}H_3$, CO$\underline{C}H_3$	*c.* 17–20
Isopropenyl (-$CH_2CH{=}C(CH_3)_2$)	21(CH_2), 122(CH), 131(C), 18(CH_3)

[1] ^{13}C-^{13}C coupling is not observed.

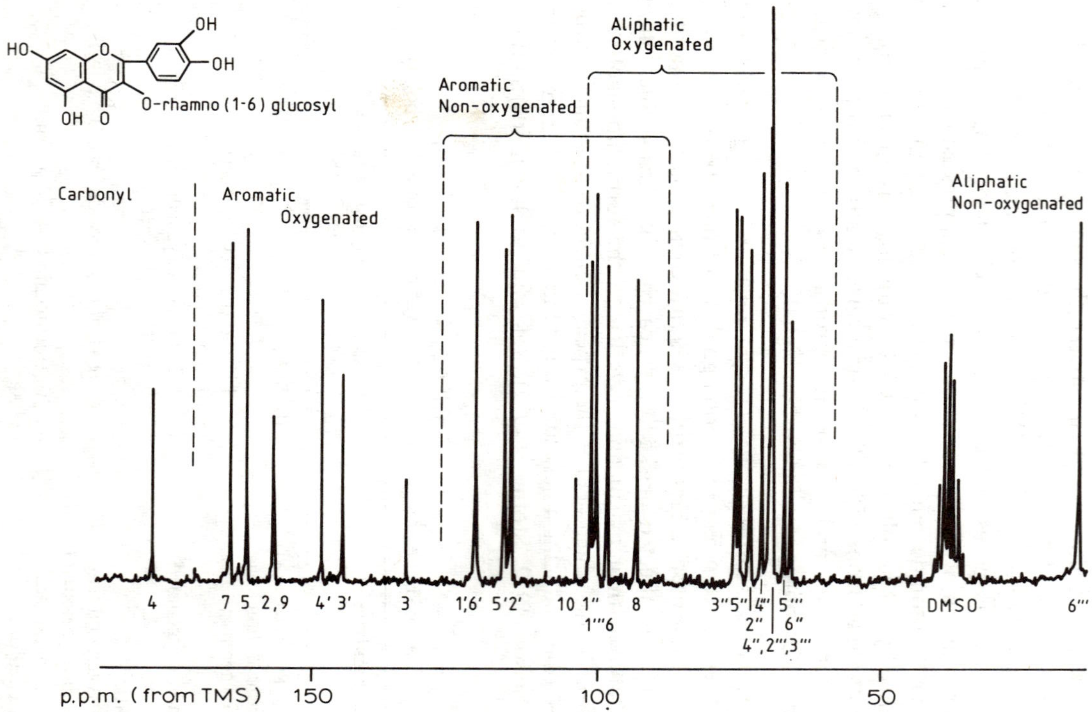

Fig. 6.1. ^{13}C-n.m.r. spectrum of rutin.

for the effect of new substituents at the "C-1", *ortho*-, *meta*- and *para*-positions (in p.p.m.) are as follows:

	C-1	*Ortho*	*Meta*	*Para*
Hydroxyl	+26·9	−12·7	+1·4	−7·3
Methoxyl	+31·4	−14·4	+1·0	−7·7
Methyl	+ 8·9	+ 0·7	−0·1	−2·9
Acetoxyl	+23·0	−6·4	+1·6	−2·3

Using this type of substituent effect data, it is possible to calculate with some accuracy the spectrum of an unknown flavonoid from that of a similarly substituted known. To do this it is necessary to have available a wide range of reference spectra and a number of such compilations have appeared, e.g. Wagner *et al.* (1976); Pelter *et al.* (1976); Markham *et al.* (1978) and Mabry *et al.* (1982) and see Table 6.4.

Identification of sugars Although ^{13}C-n.m.r. data are undoubtedly useful for defining the oxygenation patterns of flavonoid unknowns (cf. Fig. 6.4 and Spectra 7 and 8 for a varety of B-ring patterns), it is perhaps of most value in the definition of the sugar moiety and its substituents. This is especially true for flavonoid *C*-glycosides, the sugars of which are difficult to

Table 6.3

Sugar carbon resonances in the ^{13}C-n.m.r. spectra of some flavonoid *C*- and *O*-glycosides

Glycoside type[a, b]	C-1	C-2	C-3	C-4	C-5	C-6
O-glucoside (7-linked)	100·2	73·3	76·6[c]	69·8	77·4[c]	60·9
C-glucoside (8-linked)	73·9	71·4	78·8	70·8	81·4	61·5
O-galactoside (3-linked)	102·3	71·3	73·4	68·0	75·8	60·8
C-galactoside (8-linked)	73·9	68·5	75·4	69·1	80·5	61·3
O-α-rhamnoside (3-linked)	101·9	70·4[c]	70·6[c]	71·5	70·1[c]	17·3
C-α-rhamnoside (8-linked)	77·3	75·0	75·5	72·2	72·2	18·1
O-xyloside (2″-*O*-linked)	102·4	73·7	75·9	69·4	65·5	
C-xyloside (6-linked)	74·6	70·3	78·5	70·0	70·0	
O-α-arabinoside (3-linked)	101·8	71·7	70·8	65·9	64·1	
O-α-arabinofuranoside (3-linked)	108·1	82·1	77·2	86·2	61·0	
C-α-arabinoside (8-linked)	74·4	68·2	74·5	68·9	71·0	
O-glucuronide (3-linked)	101·1	73·7	75·9	71·3	75·9	169·7
O-alloside (3-linked)	99·9	71·6	71·6	67·2	75·1	61·3
O-apioside (2″-*O*-linked)	109·0	76·5	79·1	74·0	64·4	

[a] β-Linked pyranosides unless otherwise stated; solvent d_6-DMSO.
[b] Linkage site may affect C-1 resonances of *O*-glycosides (usually ±2 p.p.m.), and other resonances in *C*-glycosides (usually ±0·5 p.p.m.).
[c] Assignments may be reserved.

identify by any other method. All commonly encountered sugars, whether they be *O*- or *C*-linked give different patterns of carbon-13 resonance signals (see listing in Table 6.3, the data for which are derived mainly from Markham and Chari, 1982). Additional substituents on these sugars are readily identified from the ^{13}C-n.m.r. spectrum, as also are the points of attachment to the sugar (from the resultant signal shifts). Specific examples are given in section 6.2.4 but in general terms sugar-related substitution-induced shifts are as follows (Chari *et al.*, 1977; Markham *et al.*, 1978):

i. *O-Glucosylation of a flavonoid* hydroxyl produces an upfield shift[1] of up to 2 p.p.m. in the signal of the adjacent flavonoid carbon and downfield shifts in the signals of the *ortho*- and especially *para*-related carbons (1–4 p.p.m.).
ii. *C-Glycosylation of a flavonoid* aglycone causes a 10 p.p.m. downfield shift of the signal of the glycosylated carbon, but leaves other signals relatively unaffected.
iii. *Glycosylation of a sugar* by glucose causes an 8 p.p.m. downfield shift in the signal of the glycosylated carbon accompanied by a 1–3 p.p.m. upfield shift in the signals of adjacent carbons. Glycosylation by rhamnose causes smaller shifts (3–6 p.p.m. downfield and 1–2 p.p.m. upfield respectively).
iv. *Acylation* (e.g. acetates, benzoates etc.) of a sugar generally produces a downfield shift of about 2 p.p.m. in the signal of the acylated carbon, accompanied by upfield shifts of 1–3 p.p.m. in the signals of adjacent carbons (see section 6.2.4).

6.2.3 *Sample size and solvents*

With regard to sample size, an important principle to remember is that the smaller the sample the longer the instrument time involved in obtaining a reasonable spectrum. Thus, although with modern Fourier Transform equipped spectrometers samples of 2–5 mg are acceptable, samples of 10–50 mg (*c*. 0·1 mmol) are preferred.

The solvent most frequently used with flavonoids is d_6-DMSO and the bulk of the reference data relates to d_6-DMSO solutions. This solvent, however, produces a symmetrical pattern of seven signals centred at 39·5 p.p.m. (see Fig. 6.1), and if these obscure flavonoid carbon signals the spectrum can be rerun in $(CD_3)_2CO$ or pyridine-d_5.

6.2.4 *Example spectra and their interpretation*

In the authors view, the most useful application of ^{13}C-n.m.r. spectroscopy in flavonoid work is in the elucidation of structural features of the saccharide

[1] i.e. to lower number!

portion of a flavonoid glycoside. Examples in this section are therefore confined to those which highlight aspects of this application. The reader is referred to the work of Pelter *et al.* (1976), Wagner *et al.* (1976) and Markham and Ternai (1976) for applications of ^{13}C-n.m.r. data to structural problems involving the flavonoid nucleus, although a selection of data from these and other sources is presented in Table 6.4.

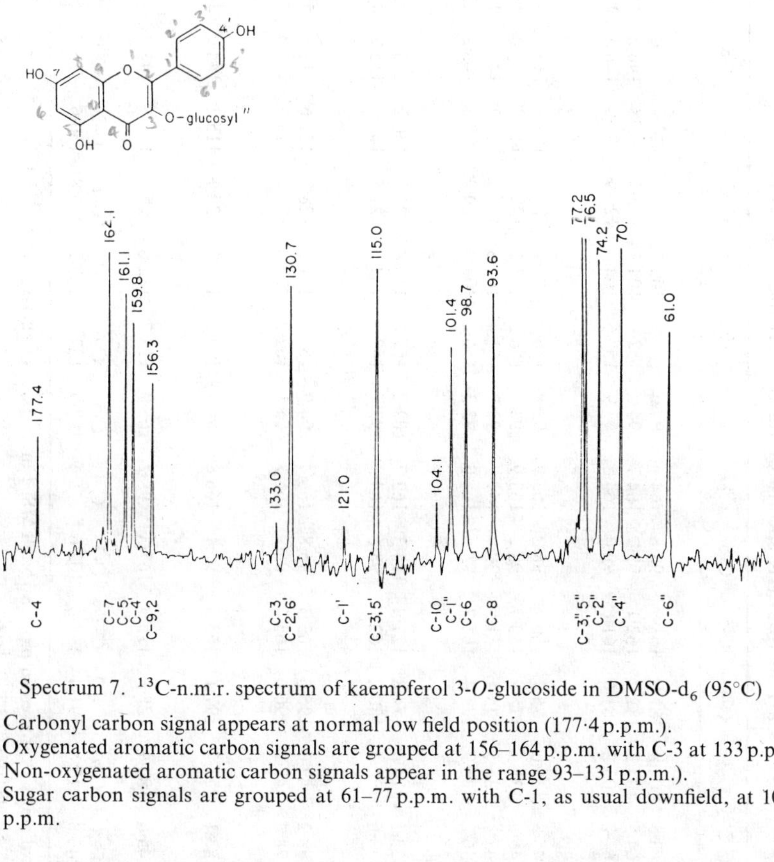

Spectrum 7. ^{13}C-n.m.r. spectrum of kaempferol 3-*O*-glucoside in DMSO-d_6 (95°C)

1. Carbonyl carbon signal appears at normal low field position (177·4 p.p.m.).
2. Oxygenated aromatic carbon signals are grouped at 156–164 p.p.m. with C-3 at 133 p.p.m.
3. Non-oxygenated aromatic carbon signals appear in the range 93–131 p.p.m.).
4. Sugar carbon signals are grouped at 61–77 p.p.m. with C-1, as usual downfield, at 101·4 p.p.m.

The effect of acylation: Kaempferol 3-*O*-(6″-*O*-acetyl) glucoside. The spectrum of this compound (not illustrated) differs from spectrum 7 in the following ways:

1. An extra carboynl carbon signal at 169·5 p.p.m. and a methyl signal at 19·9 p.p.m. represent the additional acetyl function.
2. The signal of the glucose C-6″ is shifted downfield from 61·0 to 62·9 p.p.m. and the signal of the (adjacent) C-5″ has shifted upfield from *c.* 77 p.p.m. to 74 p.p.m., due to the acetylation.

Table 6.4
Carbon-13 NMR spectra of various flavonoid aglycones[a]

Flavonoid	C-2	C-3	C-4	C-5	C-6	C-7	C-8	C-9	C-10	C-1′	C-2′	C-3′	C-4′	C-5′	C-6′	CH_3
Flavones																
Apigenin	163·8	102·8	181·8	161·9	98·8	164·1	94·0	157·3	103·7	121·3	128·4	116·0	161·5	116·0	128·4	
Luteolin	164·5[c]	103·3	182·2	162·1	99·2	164·7[c]	94·2	157·9	104·2	122·1	113·8	146·2	150·1	116·4	119·3	
Tricetin	164·2	103·2	181·6	161·6	99·0	164·2	93·9	157·5	104·0	120·9	106·0	146·5	137·9	146·5	106·0	
Flavonols																
Kaempferol	146·8	135·6	175·9	160·7	98·2	163·9	93·5	156·2	103·1	121·7	129·5	115·4	159·2	115·4	129·5	
Quercetin	146·9	135·5	175·8	160·7	98·2	163·9	93·3	156·2	103·1	122·1	115·3[c]	145·0	147·6	115·6[c]	120·0	
Patuletin	147·1[d]	135·5	176·1	151·8[c]	130·9	157·2	93·7	151·4[c]	103·5	122·1	115·2[e]	145·1	147·8[d]	115·7[e]	120·1	60·3
Myricetin (3-*O*-galactose)	156·2	133·9	177·4	161·2	98·6	164·0	93·3	156·2	104·0	120·2	108·8	145·3	136·6	145·3	108·8	
Isoflavones																
Formononetin	153·1	124·4	174·8	127·4	115·3	162·7	102·2	157·6	116·8	123·3	130·2	113·7	159·1	113·7	130·2	55·2
Genistein	153·6	122·4	180·2	162·1	98·6	164·3	93·7	157·5[c]	104·6	121·4	130·0	115·2	157·6[c]	115·2	130·0	
Dihydroflavones																
Naringenin	78·2	42·0	195·6	163·3	95·8	166·4	94·9	162·7	101·9	128·9	127·6	115·1	157·4	115·1	127·6	
Eriodictyol	78·4	42·2	196·0	163·6	95·9	166·7	95·1	162·9	102·0	129·7	114·5[d]	145·3[c]	145·7[c]	115·6[d]	117·9	
Dihydroflavonol																
Dihydroquercetin	83·1	71·7	197·1	163·3	96·1	166·8	95·1	162·5	100·6	128·1	115·3	144·9[c]	145·7[c]	115·3	119·2	
									-CH=							
Aurone																
Sulphuretin[b]	145·6	180·9	125·3	115·9	167·3	98·2	165·9	113·7	112·6[c]	123·3	111·5[c]	145·3	147·7	117·9	124·2	
Chalcone	C-β	C-α	C=0	C-2′	C-3′	C-4′	C-5′	C-6′	C-1′	C-1	C-2	C-3	C-4	C-5	C-6	
Isoliquiritigenin[b]	143·8	117·8	191·4	164·6[c]	102·6	165·4[c]	107·9	132·3	113·2	125·8	130·6	115·8	159·9	115·8	130·6	

[a] *Solvent*: DMSO-d_6, chemical shifts expressed in p.p.m. from TMS. For carbon numbering see Fig. 1.1.
[b] *Note*: different numbering systems (refer to Fig. 1.1).
[c,d,e] Assignments bearing the same superscript in any one spectrum may be reversed.

The effect of O-glucosylation: Kaempferol 3-*O*-(gluco (1–2) glucoside). Relative to spectrum 7:

1. The sugar carbon region becomes more complex with some doubling of signals.
2. A new sugar C-1 signal is evident at 103·6 p.p.m.
3. A "new" signal appears at 82·0 p.p.m. (C-2″ which has shifted downfield from 74·2 p.p.m. as a result of 2″-*O*-glucosylation).

The effect of O-rhamnosylation: Kaempferol 3-*O*-(rhamno (1–6) glucoside). Relative to spectrum 7:

1. A methyl signal (17·7 p.p.m.) and five additional sugar carbon signals at: 100·6 (C-1‴), 70·3/70·7 (C-2‴, 3‴) 72·0 (C-4‴) and 68·1 (C-5‴) appear (due to thamnose).
2. The glucose C-6″ signal shifts downfield from 61·0 to 67·1 p.p.m. (due to rhamnosylation at C-6″).

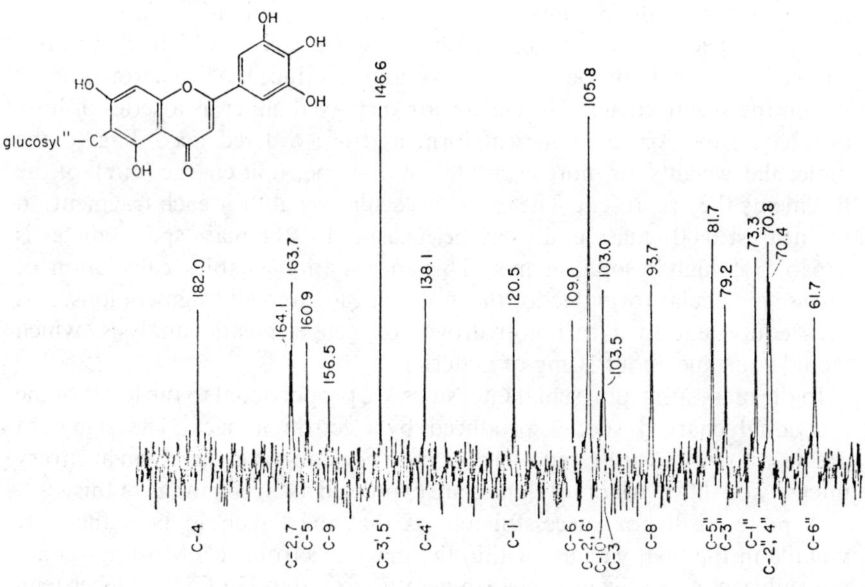

Spectrum 8. ^{13}C-n.m.r. spectrum of tricetin 6-*C*-glucopyranoside (Isoaffinetin) in DMSO-d_6 (30°C)

1. The presence of a *C*-linked substituent at C-6 is evident from the shift of the C-6 signal to 109 p.p.m. (from its normal position of *c.* 98 p.p.m. in 5,7-dihydroxyflavonoids—see spectrum 7).
2. The pattern of sugar carbon signals identifies the sugar as glucose (see Table 6.3).
3. In contrast to an *O*-linked glycoside, the signals of all *six* sugar carbons appear in the 60–82 p.p.m. range.

6.3 Mass spectroscopy

6.3.1 *Typical applications*

i. Determination of molecular weights.
ii. Establishment of the distribution of substituents between the A- and B-rings.
iii. Determination of the nature, and site of attachment, of the sugar(s) in flavonoid C- and O-glycosides.

6.3.2 *Introduction to mass spectroscopy (m.s.)*

The main advantage that m.s. has over most other physical techniques is that only sub-milligram sized samples are required for analysis. Thus, although m.s. frequently produces information which can also be obtained by other techniques, e.g. ^{13}C-n.m.r. spectroscopy, m.s. may well be the only method suitable when small amounts of the unknown are available.

A mass spectrum consists of a series of signals each of which represents a charged fragment of the parent flavonoid produced by electron impact within the spectrometer. The signals are displayed either as a series of lines on chart paper or in numerical form and are ordered according to the molecular weights, or more accurately mol.wt per unit charge (m/z), of the fragments they represent. The exact molecular weight for each fragment, to the nearest 0·001 mass unit, may be measured if the mass spectrometer is capable of high resolution m.s. This information enables calculation of precise molecular formulae for the molecular ion and all fragment ions. It is thus equivalent to a carbon/hydrogen/oxygen elemental analysis, which would consume about 10 mg of material.

In the mass spectrum, signal intensities are proportional to the levels of the individual charged species produced by electron impact. The signal of greatest intensity is referred to as the *base peak* and is assigned an arbitrary intensity of 100%. All other signals are expressed as a percentage of this.

A prerequisite for successful m.s. is that the flavonoid be sufficiently volatile in the high vacuum within the mass spectrometer. Most aglycones are sufficiently volatile at probe temperatures of 100–230°C, the higher temperatures being required for the more polar polyhydroxy-flavones and flavonols. Glycosides, anthocyanidins and biflavonoids, however, are not sufficiently volatile and must therefore be derivatized to improve their volatility. The standard method is that of permethylation, or perdeuteromethylation (for procedure see Chapter 5) although trimethylsilylation has also been used occasionally, e.g. for glycosides (Schels *et al.*, 1977, 1978) and anthocyanidins (Bombardelli *et al.*, 1977). Derivatization might also be necessary for other aglycones if they prove to be insufficiently volatile, or if it is desired to carry out purification by gas chromatography prior to m.s. (e.g. Schmid *et al.*, 1973).

Fragmentation of flavonoids and their glycosides on electron impact occurs in a limited number of predictable ways (see review, Mabry and Markham, 1975, and Pelter *et al.*, 1965). A brief summary of the more frequently encountered fragmentations is presented below, first for aglycones and then for glycosides.

6.3.3 *Interpretation of flavonoid aglycone mass spectra*

The first objective in interpreting a flavonoid m.s. is to identify the unfragmented molecular ion (M^+), and then to relate other major fragments to it by rationalizing the loss in molecular weight using recognized fragmentation pathways. Commonly encountered fragments resulting from such pathways are listed below.[1]

Molecular ion (M^+) The molecular ion normally appears as a major peak in the m.s. of aglycones, and exact mass measurement of this permits calculation of the elemental composition of the aglycone. Recognition of M^+ is aided by the fact that it must be an even mass number. Further, it must represent a reasonable molecular weight bearing in mind the mol.wt. of the basic flavonoid nucleus, i.e. flavones, isoflavones and aurones, 222; flavanones and chalcones, 224; flavonols, 238; and dihydroflavonols, 240; and the 16 mass units which must be added for each —OH, the 30 m.u. for each —OCH_3 and the 33 m.u. for each —OCD_3.

M^+-1 Loss of hydrogen. Common in most flavonoid types.

M^+-15 If the flavonoid contains a methoxyl at C-6 or C-8, the loss of $CH_3^{\cdot}$ (a radical) produces an intense ion (M^+-15, see Scheme 6.1) which may be more intense than M^+. 6- and 8-Methoxy flavonoids can be clearly distinguished from one another by the relative intensities of these and other ions (Goudard *et al.*, 1978, 1979). Lower intensity M-15 ions can originate from flavonoids *O*-methylated at other sites.

M^+-17 Loss of $OH^{\cdot}$ is indicated by the presence of an M-17 ion. This usually involves internal ring formation and is commonly associated with 2′-hydroxylation in flavones, flavonols, isoflavones etc. A loss of 31 m.u. may

[1] Beware of a major ion at m/z 149. This often indicates the presence of phthalate plasticizers (contaminants which originate from solvents, plastics etc). Lower intensity ions from the same impurities commonly occur at m/z 167 or 177.

indicate loss of $OCH_3^{\cdot}$ from a 2′-methoxylated flavonoid by the same mechanism.

M^+-18 Loss of H_2O. Common in flavonols, flavan 3,4-diols and *C*-glycosides.

M^+-28(29) Indicates loss of CO (CHO) from the 4-keto function to form a 5-membered ring (especially with *3-hydroxy*flavones and dihydroflavonols, e.g. Scheme 6.1).

M^+-31 See M-17 above.

M^+-43 The loss of CH_3 in concert with CO (i.e. CH_3CO) may also occur in methoxylated flavonoids (e.g. Scheme 6.1) leading to a loss of 43 m.u.

M^+-55(56) These ions may indicate the presence of an isopentenyl (-CH_2-$CH{=}C(CH_3)_2$) substituent and relate to the loss of a portion of this substituent.

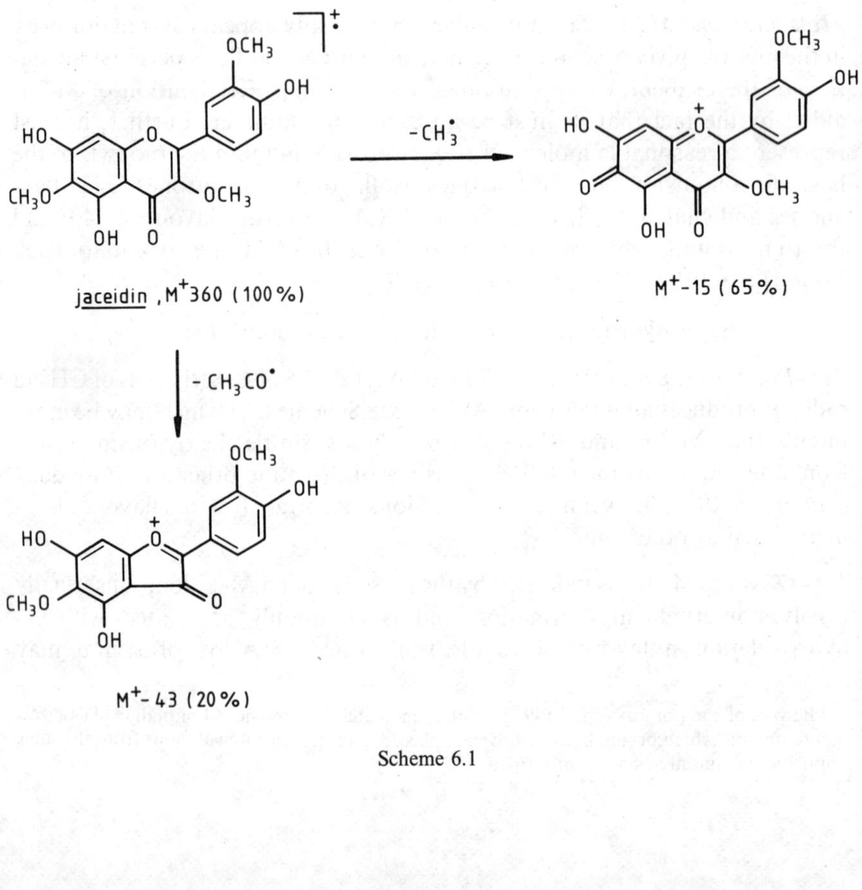

Scheme 6.1

A- and B-ring fragments Fission of the M^+ ion into A- and B-ring containing fragments often provides structurally useful information. These fragmentations usually involve one of the two competing pathways (Scheme 6.2), I (retro-Diels Alder and II. The dominant pathway is determined by the

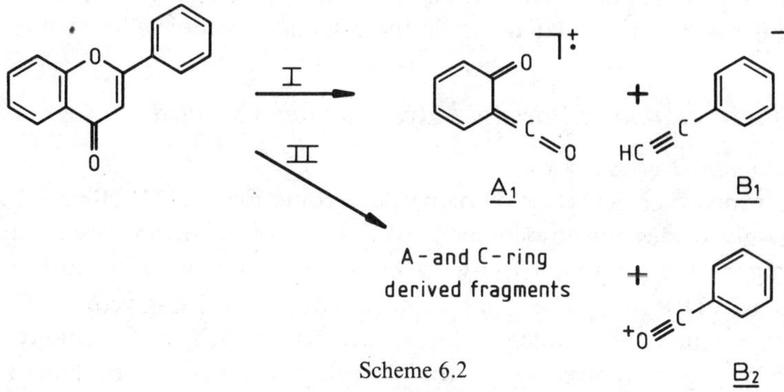

Scheme 6.2

aglycone type, although on occasions neither pathway produces detectable fragments. Flavones and isoflavones tend to produce A_1^{+} $(A_1+H)^+$ plus B_1^{+} fragmets (e.g. Scheme 6.3); flavonols, $(A_1+H)^+$ plus B_2^{+} fragments; flavanones, A_1^{+}, $(A_1+H)^+$ plus $(B_1+2H)^+$ fragments and dihydroflavonols, $A_1{}^+$ plus $(B_1+H_2O)^+$ fragments. Chalcones tend to produce fragments derived from cleavage either side of the carbonyl, although 2′-hydroxychalcones may isomerize to flavanones and produce typical flavanone fragments instead. *O*-linked biflavonoids undergo fission on either side of the interflavonoid bond and the individuai flavonoids so produced, undergo fragmentation as above. Each flavonoid unit in *C*-linked biflavonoids fragments in the normal (above) way, but due to the presence of the stable interflavonoid C–C bond, some fragments from each flavonoid remain linked to one another. Thus fragments such as (A+flavonoid)$^+$, (B+flavonoid)$^+$, (A+A)$^+$ and (A+B)$^+$ are observed, depending upon the position of the interflavonoid linkage.

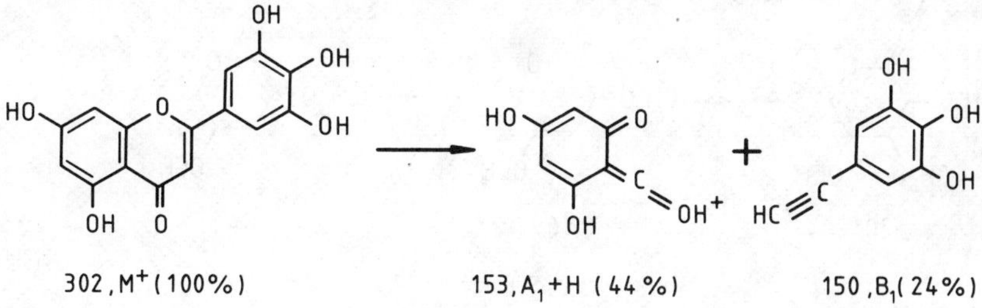

Scheme 6.3

It is clear from the example in Scheme 6.3 that the ring in which a methoxyl is sited would be readily determined from the mol.wt. of the fragments. Similarly, if the site of glycosylation is sought, the glycoside is first permethylated (or perdeuteromethylated if methoxyls are already present), and then acid hydrolysed before the m.s. is run. The fragment with the free —OH thus identifies the glycosylated ring in the original flavonoid.

6.3.4 *Interpretation of flavonoid glycoside mass spectra*

a. Flavonoid O-glycosides

Although most permethyl (PM) or perdeuteromethyl (PDM) ethers of 3,5 and 4′-*O*-glycosides give low-intensity (0·1–2%) molecular ions due to rapid loss of sugar[1], the equivalent 7-*O*-glycosides normally give molecular ions of high intensity (10–90%). The sugar fragments from mono-*O*-glycosides (PM) are produced mainly by sequential loss of MeOH (see Scheme 6.4) and reveal the type of sugar involved. Thus for permethylated derivatives, hexosides produce ions at m/z 218, 187, 155 and 111; glucuronides, 201, 169, 141 and 111; rhamnosides, 189, 157, 125 and 111, and pentosides, 175, 143 and 111. Such data are often of use in determining the site of additional glycosylation, ethylation etc. (i.e. -OCH_3 replaced by -OH or OCH_2CH_3). In di- or tri-

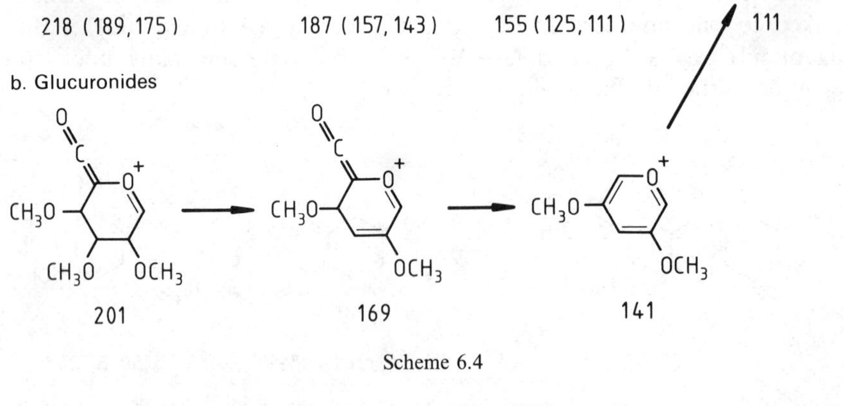

Scheme 6.4

[1] Unless chemical ionization m.s. is used—see Foltz (1972).

glycosides, sequential loss of individual sugars is normal (Schmid, 1972; Schmid *et al.*, 1972; Seligmann and Wagner, 1978). From the nature of the fragments detected it is possible to determine the sugar sequence, the positions of interglycosidic linkages and the ring size of the sugars. The use of PM-isopropylidene derivates (in which pairs of spacially close hydroxyl groups are first bridged by isopropylidene formation, see method in Chapter 5, and then permethylated) has been used with success in this work (Seligmann and Wagner, 1978).

b. Flavonoid C-glycosides

Mass spectroscopy of underivatized flavonoid *C*-glycosides rarely produces a molecular ion and gives little information about the sugar moiety. However, it does produce a base peak ion which consists of the aglycone fragment containing a CH_2 remnant of *each* original *C*-linked sugar (see Scheme 6.5 and Oelrichs *et al.*, 1968). The mol. wt of this ion is useful in defining the aglycone

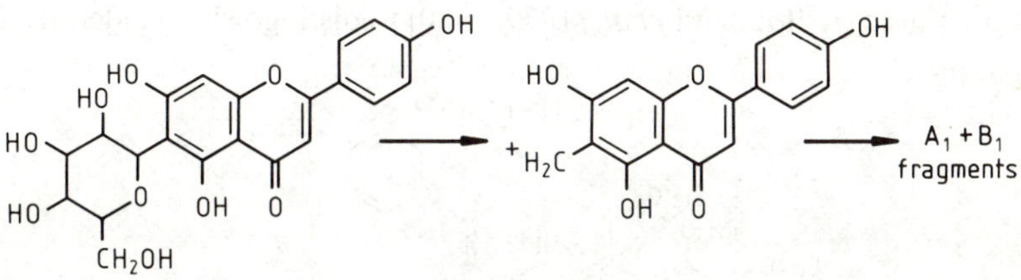

Scheme 6.5

type, as also is its subsequent fragmentation into A- and B-ring fragments (as described above for aglycones).

The PM-mono- and di-*C*-glycosides of flavonoids give well-defined mass spectra including a sizeable M^+ ion (15–100%, Bouillant *et al.*, 1975). If the aglycone moiety is known, the M.W. of the M^+ ion defines the type(s) of sugar(s) linked to the flavonoid. The list below for PM-apigenin *C*-glycosides exemplifies this use and the figures quoted can be modified to apply to any aglycone type by simply adding 30 m.u. for each additional —OH or —OCH_3.

486 m.u.	6- or 8-mono-*C*-pentosylapigenin
500 m.u.	6- or 8-mono-*C*-rhamnosylapigenin
530 m.u.	6- or 8-mono-*C*-hexosylapigenin
660 m.u.	6,8-di-*C*-pentosylapigenin
704 m.u.	*C*-pentosyl-*C*-hexosylapigenin
718 m.u.	*C*-rhamnosyl-*C*-hexosylapigenin
748 m.u.	6,8-di-*C*-hexosylapigenin

The 6-*C*-glycosides are readily distinguished from their 8-*C*-glycoside isomers by the presence in their spectra of an intense M-31 ion, which is absent from the spectra of the 8-*C*-glycosides. This same ion is also present in the spectra of 6,8-di-*C*-glycosides and is generally the base peak.

In the m.s. of 6,8-di-*C*-glycosyl flavones containing a hexose and a pentose the sugar bonded to C-6 is indicated by the relative intensities of the M^+-131 (pentose lost) and M^+-175 (hexose lost) ions. Thus if M^+-131 > M^+-175 the pentose is at C-6, and if M^+-175 > M^+-131 the hexose is at C-6. Di-*C*-glycosides containing a hexose and rhamnose behave similarly, M^+-145 (rhamnose lost) being more intense than M^+-175 when rhamnose is at C-6. Arabinose at C-6 can be distinguished from xylose at C-6 by the ratio of the M^+-145, M^+-131 and M^+-119 ions. In PM 6-*C*-arabinosides M^+-131 > M^+-119 > M^+-145, whereas in 6-*C*-xylosides M^+-119 > M^+-131 > M^+-145 (Bouillant *et al.*, 1975).

Fragmentation pathways to account for all of the above ions have been elucidated (Bouillant *et al.*, 1978) and the structures of the fragments are detailed in Scheme 6.6 for 6-*C*-pentosylapigenin. These fragmentation data have been used by Bouillant *et al.* (1979, 1980) to distinguish and identify 2″,

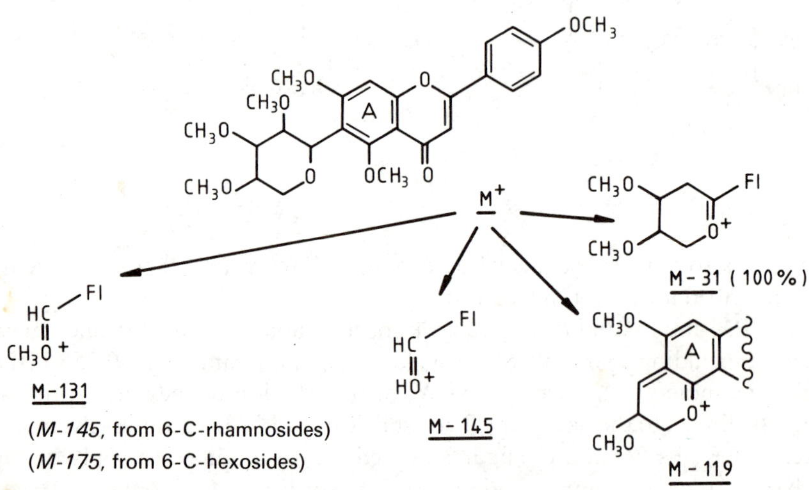

Scheme 6.6

3″,4″ and 6″-*O*-glycosides of 6-*C*-glucosylflavones. Wagner *et al.* (1979) found that formation of isopropylidene derivatives (see preparation method, Chapter 5) prior to permethylation and m.s. measurement, assisted interpretation of the spectra of *O*-glycosylated *C*-glycosylflavones.

6.3.5 *Field desorption mass spectrometry (f.d.m.s.)*

Although as yet not widely available, f.d.m.s. (Beckey and Schulten, 1975) offers a method of obtaining good quality mass spectra from *underivatized* glycosides. The technique has proved successful with a wide range of mono-, di- and tri-glycosides (molecular weights up to 1000), giving M^+ ions with intensities of 20–100% (Geiger and Schwinger, 1980) and fragments representing sequential loss of sugars.

7

Confirmation of Flavonoid Structure by Direct Comparison

Systematic application of the procedures outlined in this book should produce a structure for any unknown flavonoid. The confidence one can place in the assigned structure however, will be dependent upon the variety and quality of data supporting it, and this can vary widely. In most cases the flavonoid or its glycoside will have been reported previously, in which case published data is available for comparison. Even if the flavonoid isolated is new, it may still be possible to transform it into one that is known, e.g. by methylation, isomerization, hydrolysis etc. Agreement between published and observed data is clearly important if the assigned structure is to be published. However, physical data obtained in different laboratories for the same compound can vary, and this is especially true of melting points and R_f values and to a lesser extent of ^{13}C-n.m.r. and m.s. data. For this reason, the preferred method for confirmation of structure is *direct* comparison of the isolated flavonoid with an authentic sample from another source.

To carry out such a comparison, it is first necessary to obtain the authentic sample, and a number of avenues are open to the researcher:

i. The required flavonoid (or one that can be converted to it) may be purchased from a commercial source. A number of firms market flavonoids and a summary of some of the more useful sources (but not necessarily the only or even the best) known to the author are listed in Table 7.1.

ii. A small sample of the flavonoid can usually be obtained from the workers who reported its earlier isolation.

iii. The authentic flavonoid may be isolated from a plant source, either that reported by the original workers, or alternatively from a source referred to in review articles such as Geissman (1962), Harborne *et al*. (1975) or Harborne and Mabry (1982). Refer also to Table 1.1, Chapter 1.

Table 7.1
Some commercial sources of useful flavonoids

Flavonoid	Source[a]
Flavones	
Flavone	F,R
Chrysin (and 7-methyl ether)	R,S
7-Hydroxy-4′-methoxyflavone (pratol)	S
Apigenin	F,R,S
Acacetin	S
Acacetin 7-*O*-rutinoside	S
Apigenin 7-*O*-glucoside	R,S
Apigenin 7-*O*-apioglucoside	R,S
Apigenin 7-*O*-neohesperidoside	S,R
Apigenin 7-*O*-rutinoside	S
Luteolin	S,R
Luteolin 5-*O*-glucoside	R
Luteolin 7-*O*-glucoside	R,S
Chrysoeriol and diosmetin	S
Tricin	R
Apigenin 8-*C*-glucoside (vitexin)	R,S
Vitexin 4′-*O*-rhamnoside	R,S
Flavonols	
Galangin	S,R,F
Fisetin	S,K-L
Robinetin	S,R
Robinin	R,K-L,S
Kaempferol	S,F,Si
Kaempferol 4′-methyl ether	S,F
Kaempferol 3-*O*-rhamnoglucoside	R,S
Datiscetin	S
Quercetin	R,S,F,K-L,Si
Quercetin 3-*O*-galactoside	R,S
Quercetin 3-*O*-rhamnoside	F,S
Quercetin 3-*O*-rutinoside	F,R,K-L,Si
Quercetin 3-*O*-glucoside	S
Quercetin 3-*O*-arabinoglucoside	S
Morin	S
Tamarixetin	S
Rhamnetin	S
Isorhamnetin	S
Quercetagetin	R,S
Myricetin	F,S,Si
Myricetin 3-*O*-rhamnoside	R

Table 7.1 (cont.)

Flavonoid	Source[a]
Isoflavones	
Daidzein (and formononetin)	S
Genistein	R,S
Biochanin A	S
Chalcones	
2′,4′,6′,3,4-Pentahydroxychalcone	S
Okanin 4′-*O*-glucoside	S
"Hesperidin methyl chalcone"	K-L,Si
Butein	S
Homobutein	S
Marein	S
Neohesperidin dihydrochalcone	Si
Phloretin (and 4-methyl ether)	S
Biflavonoids	
Ginkgetin-Isoginkgetin	R
Sciadopytisin	R,S
Hinokiflavone	R,S
Aurone	
Maritimetin 6-glucoside	S
Surphuretin	S
Anthocyanins/anthocyanidins	
Pelargonidin chloride	R
Pelargonidin 3,5-diglucoside ("pelargonidin chloride")	R
Cyanidin chloride	R
Cyanidin 3,5-diglucoside ("cyanine chloride")	F
Delphinidin chloride	F,R
Malvidin 3,5-diglucoside ("malvin chloride")	R
"*Catechins*"	
(+) and (−) catechins	R,K-L,F
Epicatechin (+) and (±)	R,K-L,F
Flavanones	
Flavanone	K-L,S,
Naringenin	R,K-L,Si,S
Naringenin 7-*O*-neohesperidoside	F,K-L,R,Si,S
Hesperetin	R,Si,S
Hesperidin	R,K-L,Si,S
Eriodictyol	S
Eriodictyol 7-*O*-glucoside (and 7-rhamnoglucosides)	S
Homoeriodictyol	S

Flavonoid	Source[a]
Isookanin 7-*O*-glucoside	S
Isosakuranetin (and 7-rhamnoglucosides)	S
Dihydroflavonols	
Dihydrofisetin (fustin)	R,K-L,S
Dihydrorobinetin	R,S
(+)-Dihydroquercetin (taxifolin)	K-L,F,Si,R,S

[a] Sources: **K-L**, Koch-Light Laboratories Ltd, Colnbrook SL3 0BZ Buckinghamshire, England. **Si**, Sigma Chemical Co., PO Box 14508, St Louis, Missouri 63178, USA. **R**, Carl Roth, Schoemperlenstrasse 1–5, Postfach 21 0980, D7500 Karlsruhe 21, W. Germany. **F**, Fluka AG, Chemische Fabrik, CH-9470 Buchs, Switzerland. **S**, Sarsyntex, Laboratories Sarget B.P. 100, Av. du President Kennedy, 33701 Merignac, France.

iv. The required flavonoid may be synthesized, either by transformation of an available flavonoid (see e.g. Chapters 4 and 5) or from non-flavonoid precursors (see Geissman, 1962).

Once the authentic flavonoid has been obtained, the question arises as to how best to compare it with the flavonoid under study. Clearly, in general terms, the greater the number of techniques used, the greater will be the level of confidence in the result. In practice, however, constraints such as the amount of material and type of equipment available will limit the possibilities. *An absolute minimum* for effective comparison of two flavonoids, in the author's opinion, is cochromatography in three different systems. By cochromatography is meant the chromatographing of the two flavonoids side by side on the same chromatogram, both separately and mixed. The three different systems (solvent/adsorbent combinations) must involve three markedly different solvents and at least two different chromatographic media, e.g. cellulose, silica or polyamide. With t.l.c. or p.c. comparisons the use of different spray reagents on each of the three chromatograms will also aid in the recognition of identity. Electrophoresis, gas–liquid chromatography and high-performance liquid chromatography, if used, would each be considered additional chromatographic systems.

The comparison need not be confined solely to chromatographic comparison. Indeed, if sufficient flavonoid is available, techniques such as u.v.-visible and (especially) infrared absorption spectroscopy, ^{1}H or ^{13}C-n.m.r. spectroscopy and mass spectrometry may also be used.

If the flavonoids are crystalline, one of the best methods of comparison is by melting point and mixed melting point determinations. Identical pure (i.e. recrystallized) flavonoids will possess the same or closely similar melting points, as also will an intimately ground mixture of both. However, if the flavonoids are not identical the melting points may or may not be the same, but the melting point of the mixture will be markedly depressed, often 20–30°C below the melting point of either component. If a standard melting point apparatus is not available, melting points may be measured in a stirred, heated, oil-bath, the samples (in separate capillary tubes with sealed ends) being positioned in close proximity to one another and to the tip of the thermometer. The bath is heated with a hot-plate or Bunsen burner, but care should be taken to ensure that the rate of temperature rise is not greater than about 2°C per minute when approaching the melting range of the flavonoids.

It is worth noting that none of the procedures referred to above are suitable for comparing flavonoids which differ only in their stereochemistry at one carbon atom, e.g. (+) and (−)-flavanones. Such compounds (optical enantiomers, see Chapter 1) can only be distinguished by measurement of their optical rotation.

References

Adinarayana, D. and Seshadri, T. R. (1965). *Tetrahedron* **21**, 3727.
Ballantine, J. A. and Pillinger, C. T. (1967). *Tetrahedron* **23**, 1691.
Bannerjee, N. R. and Seshadri, T. R. (1954). *J. scient. ind. Res.* (India), **13B**, 598.
Barton, G. M. (1968). *J. Chromat.* **34**, 562.
Bass, R. J. (1979). *J. C. S. Chemical Communications*, 264.
Batterham, T. J. and Highet, R. J. (1964). *Aust. J. Chem.* **17**, 428.
Beckey, H. D. and Schulten, H. R. (1975). *Angew. Chem. internat. Edit.* **14**, 403.
Beirne, J. J., Carroll, N. M., O'Sullivan, W. I. and Woods, J. (1975). *Tetrahedron* **31**, 265.
Bohm, B. A. (1975). *In* "The Flavonoids" (J. B. Harborne, T. J. Mabry and H. Mabry, eds), p. 560. Chapman and Hall, London.
Bombardelli, E., Bonati, A., Gabetta, B., Martinelli, E. M., Mustich, G. and Danieli, B. (1976). *J. Chromat.* **120**, 115.
Bombardelli, E., Bonati, A., Gabetta, B., Martinelli, E. M. and Mustich, G. (1977). *J. Chromat.* **139**, 111.
Bouillant, M.-L., Favre-Bonvin, J. and Chopin, J. (1975). *Phytochemistry* **14**, 2267.
Bouillant, M.-L., Besset, A., Favre-Bonvin, J. and Chopin, J. (1978). *Phytochemistry* **17**, 527.
Bouillant, M.-L., Besset, A., Favre-Bonvin, J. and Chopin, J. (1979). *Phytochemistry* **18**, 690.
Bouillant, M.-L., Besset, A., Favre-Bonvin, J. and Chopin, J. (1980). *Phytochemistry* **19**, 1755.
Cahn, R. S. and Ingold, C. K. (1951). *J. chem. Soc.* 612.
Cahn, R. S., Ingold, C. K. and Prelog, V. (1956). *Experientia* **12**, 81.
Carles, J., Schneider, A. and Lacoste, A. M. (1958). *Bull. Soc. Chim. biol.* **40**, 221; *J. Chromat.* **1**, xxiii.
Chandler, B. V. and Harper, K. A. (1961). *Aust. J. Chem.* **14**, 586.
Chari, V. M., Wagner, H. and Neszmelyi, A. (1977). Proceedings of the 5th Hungarian Bioflavonoid Symposium, Matrafured, Hungary, p. 49.
Chawla, H., Chibber, S. S. and Seshadri, T. R. (1974). *Phytochemistry* **13**, 2301.
Chexal, K. K., Handa, B. K. and Rahman, W. (1970). *J. Chromat.* **48**, 484.
Chopin, J. and Bouillant, M.-L. (1975). *In* "The Flavonoids" (J. B. Harborne, T. J. Mabry, and H. Mabry, eds), p. 632. Chapman and Hall, London.

Chopin, J., Dellamonica, G., Besson, E., Skrzypczakowa, L., Budzianowski, J. and Mabry, T. J. (1977). *Phytochemistry* **16**, 1999.
Chopin, J., Besson, E. and Bouillant, M.-L. (1982). *In* "Advances in Flavonoid Research, 1975–1980" (J. B. Harborne and T. J. Mabry, eds). Chapman and Hall, London.
Dean, F. M. (1966). *Tetrahedron Lett.* 4153.
Dossaji, S. F., Mabry, T. J. and Wallace, J. W. (1975). *Rev. Latinoamer. Quim.* **6**, 37.
Farkas, L., Vermes, B., Nogradi, M. and Kalman, A. (1976). *Phytochemistry* **15**, 1184.
Foltz, R. L. (1972). *Lloydia* 344.
Geiger, H. (1982). *In* "Advances in Flavonoid Research, 1975–1980" (J. B. Harborne and T. J. Mabry, eds). Chapman and Hall, London.
Geiger, H. and Quinn, C. (1975). *In* "The Flavonoids (J. B. Harborne, T. J. Mabry and H. Mabry, eds), p. 692. Chapman and Hall, London.
Geiger, H. and Schwinger, G. (1980). *Phytochemistry* **19**, 897.
Geissman, T. A. (1962). "The Chemistry of Flavonoid Compounds". Pergamon Press, New York.
Geissman, T. A. and Harborne, J. B. (1956). *J. Am. chem. Soc.* **78**, 832.
Gornall, R. J., Bohm, B. A. and Dahlgren, R. (1979). *Botaniska Notiser* **132**, 1.
Goudard, M., Favre-Bonvin, J., Lebreton, P. and Chopin, J. (1978). *Phytochemistry* **17**, 145.
Goudard, M., Favre-Bonvin, J., Strelisky, J., Nogradi, M. and Chopin, J. (1979). *Phytochemistry* **18**, 186.
Grover, S. K., Jain, A. C. and Seshadri, T. R. (1964). *Tetrahedron* **20**, 555.
Hahlbrock, K. (1980). *Archs Biochem. Biophys.* **200**, 617.
Hahlbrock, K. and Grisebach, H. (1975). *In* "The Flavonoids" (J. B. Harborne, T. J. Mabry and H. Mabry, eds), p. 866. Chapman and Hall, London.
Harborne, J. B. (1958). *Biochem. J.* **70**, 22.
Harborne, J. B. (1965). *Phytochemistry* **4**, 107.
Harborne, J. B. (1967a). "Comparative Biochemistry of the Flavonoids". Academic Press, London and New York.
Harborne, J. B. (1967b). "Comparative Biochemistry of the Flavonoids", p. 30. Academic Press, London and New York.
Harborne, J. B. (1967c). "Comparative Biochemistry of the Flavonoids", p. 181. Academic Press, London and New York.
Harborne, J. B. (1973a). "Phytochemical Methods", p. 63. Chapman and Hall, London.
Harborne, J. B. (1973b). "Phytochemical Methods", p. 212. Chapman and Hall, London.
Harborne, J. B. (1975). *Phytochemistry* **14**, 1147.
Harborne, J. B. (1977). *In* "Progress in Phytochemistry" (L. Reinhold, J. B. Harborne and T. Swain, eds), Vol. 4, p. 189. Pergamon Press, Oxford.
Harborne, J. B. and Mabry, T. J. (eds) (1982). "Advances in Flavonoid Research, 1975–1980". Chapman and Hall, London.
Harborne, J. B., Mabry, T. J. and Mabry, H. (eds) (1975). "The Flavonoids". Chapman and Hall, London.
Haslam, E. (1975). *In* "The Flavonoids" (J. B. Harborne, T. J. Mabry and H. Mabry, eds), p. 560. Chapman and Hall, London.
Hillis, W. E. and Horn, D. H. S. (1965). *Aust. J. Chem.* **18**, 531.
Hörhammer, L., Wagner, H., Arndt, H. G., Kraemer, H. and Farkas, L. (1966). *Tetrahedron Lett.* 567.

Horowitz, R. M. and Jurd, L. (1961). *J. org. Chem.* **26**, 2446.

Huke, M., Gorlitzer, K. and Schenck, G. (1969). *Arch. Pharm., Berl.* **302**, 401.

Hurst, H. M. and Harborne, J. B. (1967). *Phytochemistry* **6**, 1111.

Ingham, J. L. (1979). *Z. Naturf.* **34C**, 159.

Ingham, J. L. and Markham, K. R. (1980). *Phytochemistry* **19**, 1203.

Ingham, J. L., Keen, N. T. and Hymowitz, T. (1977). *Phytochemistry* **16**, 1943.

Jackson, B., Locksley, H. D., Scheinmann, F. and Wolstenholme, W. A. (1971). *J. chem. Soc.* (C), 3791.

Jain, A. C. and Gupta, R. K. (1975). *Aust. J. Chem.* **28**, 607.

Jain, A. C., Khazanchi, R. and Kumar, A. (1978). *Tetrahedron* **34**, 3569.

Jarvis, M. C. and Duncan, H. J. (1974). *J. Chromat.* **92**, 454.

Jay, M., Gonnet, J.-F., Wollenweber, E. and Voirin, B. (1975). *Phytochemistry* **14**, 1605.

Jiracek, V. and Prochazka, Z. (1963). *In* "Paper Chromatography" (I. M. Hais and K. Macek, eds), p. 254. Academic Press, New York.

Joseph-Nathan, P., Abramo-Bruno, D. and Torres, Ma A (1981). *Phytochemistry* **20**, 313.

Jurd, L. (1962). *In* "The Chemistry of Flavonoid Compounds" (T. A. Geissman, ed.), p. 107. Pergamon Press, New York.

Jurd, L. and Horowitz, R. M. (1961). *J. org. Chem.* **26**, 2561.

Karanjgaokar, C. G., Radhakrishnan, P. V. and Venkataraman, K. (1967). *Tetrahedron Lett.* 3195.

Kingston, D. G. I. (1979). *Journal of Natural Products* **42**, 237.

Kirchner, J. G. (1967). *In* "Thin-layer Chromatography", Vol. XII, "Techniques of Organic Chemistry" (E. S. Perry and A. Weissberger, eds), pp. 556–568, Interscience Publications, New York.

Kreuzaler, F. and Hahlbrock, K. (1973). *Phytochemistry* **12**, 1149.

Kubota, T., Ichikawa, N., Matsuo, K. and Shibata, K. (1966). *Tetrahedron Lett.* 4671.

Levy, G. C. and Nelson, G. L. (1972). "Carbon-13 Nuclear Magnetic Resonance for Organic Chemists" Wiley, New York.

Litvinenko, V. I. and Maksyutina, N. P. (1965). *Khim. Prir. Soedin.* No. 6, 420.

Litvinenko, V. I. and Markov, V. A. (1969). *Khim. Prir. Soedin.* **5**, No 5, 366; *Chem. Abstr.* **72**, 1111770g.

Mabry, T. J. and Markham, K. R. (1975). *In* "The Flavonoids" (J. B. Harborne, T. J. Mabry and H. Mabry, eds), p. 78. Chapman and Hall, London.

Mabry, T. J., Markham, K. R. and Thomas, M. B. (1970a). "The Systematic Identification of Flavonoids". Springer Verlag, Berlin.

Mabry, T. J., Markham, K. R. and Thomas, M. B. (1970b). "The Systematic Identification of Flavonoids", p. 253. Springer Verlag, Berlin.

Mabry, T. J., Markham, K. R. and Chari, V. M. (1982). *In* "Advances in Flavonoid Research, 1975–80" (J. B. Harborne and T. J. Mabry, eds). Chapman and Hall, London.

Marchelli, R. and Vining, L. C. (1973). *Can. J. Biochem. Physiol.* **51**, 1624.

Markham, K. R. (1975). *In* "The Flavonoids" (J. B. Harborne, T. J. Mabry and H. Mabry, eds), p. 1. Chapman and Hall, London.

Markham, K. R. (1980). *Biochemical Systematics and Ecology*, **8**, 11.

Markham, K. R. and Chari, V. M. (1982). *In* "Advances in Flavonoid Research, 1975–80" (J. B. Harborne and T. J. Mabry, eds). Chapman and Hall, London.

Markham, K. R. and Mabry, T. J. (1975). *In* "The Flavonoids" (J. B. Harborne, T. J. Mabry and H. Mabry, eds), p. 45. Chapman and Hall, London.

Markham, K. R. and Porter, L. J. (1969). *Phytochemistry* **8**, 1777.
Markham, K. R. and Ternai, B. (1976). *Tetrahedron* 2607.
Markham, K. R. and Williams, C. A. (1980). *Phytochemistry* **19**, 2789.
Markham, K. R., Porter, L. J., Mues, R., Zinsmeister, H. D. and Brehm, B. G. (1976). *Phytochemistry* **15**, 147.
Markham, K. R., Zinsmeister, H. D. and Mues, R. (1978). *Phytochemistry* **17**, 1601.
Matsuura, S., Inuma, M., Ishikawa, K. and Kagei, K. (1978). *Chem. Pharm. Bull.* **26**, (1), 305.
Narasimhachari, N. and Seshadri, T. R. (1952). *Proc. Indian Acad. Sci.* **35A**, 202.
Niklas, K. J. and Gianassi, D. E. (1978). *Am. J. Bot.* **65**, 943.
Nilsson, E. (1973a). *Chemica Scripta* **4**, 56.
Nilsson, E. (1973b). *Chemica Scripta* **4**, 49.
Oelrichs, P., Marshall, J. T. B. and Williams, D. H. (1968). *J. chem. Soc.* (C), 941.
Okigawa, M., Kawano, N., Rahman, W. and Dhar, M. M. (1972). *Tetrahedron Lett.* 4125.
Pangon, J. F., Jay, M. and Voirin, B. (1974). *Phytochemistry* **13**, 1883.
Pasakova, J. and Munk, V. (1960). *J. Chromat.* **4**, 241.
Pelter, A. and Amenechi, P. I. (1969). *J. chem. Soc.* (C), 887.
Pelter, A., Stainton, P. and Barber, M. (1965). *J. heterocyclic Chem.* **2**, 262.
Pelter, A., Ward, R. S. and Gray, T. I. (1976). *J. chem. Soc., Perkin I*, 2475.
Pew, J. C. (1948). *J. Am. chem. Soc.* **70**, 3031.
Poole, C. F. (1977). *In* "Handbook of Derivatives for Chromatography" (K. Blau and G. King, eds), p. 152. Heyden, London.
Porter, L. J. and Markham, K. R. (1970). *Phytochemistry* **9**, 1363.
Pryde, A. and Gilbert, M. T. (1979). "Applications of High Performance Liquid Chromatography". Chapman and Hall, London.
Rodriguez, E., Carman, N. J. and Mabry, T. J. (1972). *Phytochemistry* **11**, 409.
Schels, H., Zinsmeister, H. D. and Pfleger, K. (1977). *Phytochemistry* **16**, 1019.
Schels, H., Zinsmeister, H. D., and Pfleger, K. (1978). *Phytochemistry*, **17**, 523.
Schmid, R. D., Mues, R., McReynolds, J. H., Vander Velde, G., Nakatani, N., Rodriguez, E. and Mabry, T. J. (1973). *Phytochemistry* **12**, 2765.
Seligmann, O. and Wagner, H. (1978). *Tetrahedron* **34**, 3299.
Seshadri, T. R. (1962). *In* "The Chemistry of Flavonoid Compounds" (T. A. Geissman, ed.) p. 156. Pergamon Press, Oxford.
Simpson, T. H. and Beton, J. L. (1954). *J. chem. Soc.* 4065.
Smith, H. (1972). *In* "Phytochrome" (K. Mitrakos and W. Shropshire, eds), p. 433. Academic Press, New York and London.
Stahl, E. (1969). "TLC-A Laboratory Handbook", 2nd Edition. Springer Verlag, Berlin.
Stothers, J. B. (1972). "*Carbon-13 NMR Spectroscopy*", Vol. 24, "Organic Chemistry, A Series of Monographs", Academic Press, New York and London.
Swain, T. (1975). *In* "The Flavonoids" (J. B. Harborne, T. J. Mabry and H. Mabry, eds), p. 1096. Chapman and Hall, London.
Swain, T. (1976). *In* "Chemistry and Biochemistry of Plant Pigments" (T. W. Goodwin, ed.), 2nd Edition, p. 425. Academic Press, London.
Touchstone, J. C. and Dobbins, M. F. (1978). "Practice of Thin-Layer Chromatography". John Wiley and Sons, New York.
Tyukavkina, N. A. and Pogodaeva, N. N. (1975). *Khim. Prir. Soedin*, No. 6, 708.
Vermes, B., Farkas, L., Nogradi, M., Wagner, H. and Dirscherl, R. (1976). *Phytochemistry* **15**, 1320.

Vogel, A. I. (1961). *In* "A Textbook of Practical Organic Chemistry", 3rd Edition, p. 969–972. Longmans Green & Co. London.

Wagner, H., Chari, V. M. and Sonnenbichler, J. (1976). *Tetrehedron Lett.* No. 21, 1799.

Wagner, H., Obermeier, G., Seligmann, O. and Chari, V. M. (1979). *Phytochemistry* **18**, 907.

Williams, C. A., Harborne, J. B. and Clifford, H. T. (1973). *Phytochemistry* **12**, 2417.

Wollenweber, E. and Dietz, V. H. (1981). *Phytochemistry*, **20**, 869.

Wong, E. (1976). *In* "Chemistry and Biochemistry of Plant Pigments" (T. W. Goodwin, ed.), 2nd Edition, p. 464. Academic Press, London.

Index